Progress in Geography

INTERNATIONAL REVIEWS OF CURRENT RESEARCH

Volume 2

Progress in Geography

INTERNATIONAL REVIEWS OF CURRENT RESEARCH

Volume 2

edited by

Christopher Board, *Senior Lecturer in Geography, London School of Economics*

Richard J. Chorley, *Fellow of Sidney Sussex College, Cambridge; University Lecturer in Geography*

Peter Haggett, *Professor of Urban and Regional Geography, University of Bristol*

David R. Stoddart, *Fellow of Churchill College, Cambridge*

EDWARD ARNOLD

First published 1970 by
Edward Arnold (Publishers) Ltd,
41 Maddox Street, London W1

SBN: 7131 5540 x

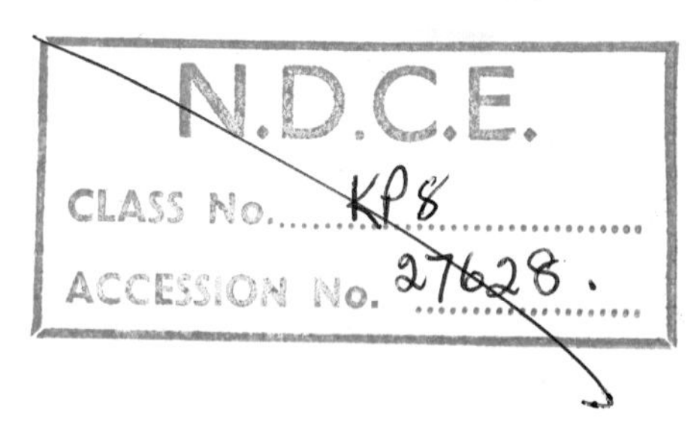

Printed in Great Britain by
Butler & Tanner Ltd, Frome and London

General preface

Events of recent years have conspired to produce an ever-rising and widening flood of geographical publication. It has been estimated that the current annual output of geographical publications is between ten and twenty thousand. The 'explosion of the geographical data matrix', the increasing popularity and relevance of geographical studies, their widening scope and their deepening analytical sophistication have been the causes of this upsurge. It has taken place, furthermore, not only in recognised geographical organs, but in a host of allied journals and, particularly, in the form of ephemeral and fugitive literature of narrow circulation. Indeed it is apparent that much of the most advanced and exciting geographical research is being reported in the first instance outside the established geographical periodicals. Surveys in Great Britain have shown that half of the literature issued by libraries is in a subject different from the field in which the user has been trained. Geography *sui generis*, because of its considerable overlaps with allied fields, suffers especially from the wide dispersal of its research literature.

The quantity of scientific literature is believed to double about every fifteen years, and there seems no reason to suppose that geographical literature is exceptional. While this incessant, logarithmic, if often obscure growth of geographical research is a welcome and long-overdue sign of intellectual vigour within the subject, it is also placing increasing strains on workers who attempt to keep abreast of developments over more than a very narrow front. To judge from the very scanty evidence provided by readership surveys and from patterns of citations, geographical publications reach only some ten to twenty per cent of the potential readership. Communication of research results is much more of a problem than it is thought to be. Not only does the specialist geographer draw much of his inspiration from other parts of the discipline, but the viability of geography as a whole is dependent upon the existence of a sympathetic corporate awareness, and this is becoming increasingly difficult to maintain on any but the most superficial level. Nor do the existing abstracting journals adequately counter these difficulties, for the significance of individual publications is only really apparent when placed within the context of larger bodies of scholarship.

Progress in Geography has been instituted with this information crisis as its central concern. Its aims are to present regular, scholarly reviews of current developments within all branches of the field on a scale which will allow the specialist contributors an opportunity to develop broad geographical themes and to provide comprehensive bibliographic material. It is hoped that contributions will be catholic in scope, taking the form both of reviews of emerging areas of new research and of stock-taking of work within more traditional frameworks. Both regional and systematic geography will be represented in traditional as well as contemporary modes of analysis. In short, the editorial objectives are to provide informed and informative reports and evaluations of whatever active research is going on within the whole field of geography.

CHRISTOPHER BOARD
RICHARD J. CHORLEY
PETER HAGGETT
DAVID R. STODDART

Advisory editors

Acknowledgements

The editors and publisher gratefully acknowledge permission given by the following to reproduce copyright material:

MIT Press and the author for figures from *The image of the city* by Kevin Lynch, 1960 ['Geographic space perception', Figs. 1–7]; the editor and the Institute of British Geographers for figures from 'A contribution to the dynamic climatology of the equatorial eastern Pacific and central America, based on meteorological satellite data' by Eric C. Barrett, *Transactions* 50, 1969 ['Rethinking climatology', Figs. 11–19, 21, 23]; Longman Group Ltd. for Fig. 12 in *Viewing weather from Space* by E. C. Barrett, 1967 ['Rethinking climatology', Fig. 8]; the editor for a figure from 'Aspects of hurricane structure: new model considerations suggested by TIROS and Project Mercury observations' by R. W. Fett, *Monthly Weather Review* 92, 43–60, 1964 ['Rethinking climatology', Fig. 7]; the author and United States Department of Commerce for a figure from 'Wind speeds from TIROS pictures of storms in the tropics' by Andrew Timchalk, *Meteorological Satellite Report* 33, 1965 ['Rethinking climatology', Fig. 9]; the author and the United States Department of the Air Force for a figure from 'Cloud interpretation from satellite altitudes' by J. H. Conover, *Cambridge Research Laboratories, Research Note* 81, Supplement 1, 1963 ['Rethinking climatology', Fig. 2]; 'Prediction of Beach Changes' by W. Harrison was published originally in *Marine Geology* 7, pp. 529–55, 1969 and is reprinted here, with some minor modifications, by permission of the Elsevier Publishing Company, Amsterdam.

Contents

General preface v

Quantitative methods in regional taxonomy 1
Nigel A. Spence, *Lecturer in Geography, London School of Economics* and
Peter J. Taylor, *Demonstrator in Geography, University of Newcastle-Upon-Tyne*

Geographic space perception: past approaches and future prospects 65
Roger M. Downs, *Assistant Professor of Geography, Pennsylvania State University*

Building models of urban growth and spatial structure 109
Robert J. Colenutt, *Assistant Professor of Geography, Syracuse University*

Rethinking climatology: an introduction to the uses of weather satellite photographic data in climatological studies 153
Eric C. Barrett, *Lecturer in Geography, University of Bristol*

Prediction of beach changes 207
W. Harrison, *The Virginia Institute of Marine Science, Gloucester Point, Virginia*

Quantitative methods in regional taxonomy

by Nigel A. Spence
and Peter J. Taylor

Contents

Preface[1]

> To see region construction, one of the last preserves of the non or anti-mathematical geographers, crumble away before the ever growing appetite of the computing machines is a little unnerving even for a hard case quantifier. (Bunge, 1966a, xix)

GIVEN a set of objects—they may be animals, plants, people, central places or almost anything of interest in a research project—then the taxonomic problem in simplest terms is to rearrange the objects into a system of classes on the basis of some measurements on the objects. The result is a classification. If the objects are areal units then the classification has produced 'regional types'. If, however, the arrangement of areal units has been carried out so as to allocate only contiguous areal units to the same class or 'region' then the solution is a regionalisation. It is this process, regionalisation, that has stimulated a large literature, mostly by geographers. The most obvious example is the work of the International Geographical Union's Commission on Methods of Economic Regionalisation.

Coincident with the writing of this perspective was the availability of this Commission's Final Report (Berry and Wróbel, 1968). Altogether an impressive list of nine publications of the Commission has been issued, (IGU Commission, 1962, 1964, 1965, 1968; Berry and Hankins, 1963; Streumann, 1967; Claval and Juillard, 1967; Macka, 1967). They discuss various aspects of economic regionalisation, including the development of basic concepts towards a general theory and typology, the role of economic regions in economic development and their practical value in administration and planning, and, of particular interest here, some of the many numerical methods of regionalisation (Dziewoński, 1968). However this latter part of the Commission's work has tended to be restricted in the range of possible approaches investigated. In particular it has not taken full advantage of the clear analogy between regionalisation and classification drawn by geographers in recent years (Berry, 1958; Bunge, 1966b, 1966c; Grigg, 1965, 1967; Johnston, 1968). In fact this similarity is so marked that Bunge (1966c) would argue for an 'isomorphic relationship' rather than an analogy. Our purpose here is to review the quantitative methods that are available for use in regionalisation and, as most of these have been developed to solve classification problems outside geography, this marked similarity between regionalisation and classification can be considered the basic assumption of this paper. However we do not intend to

[1] This paper is a revised and extended version of two previously unpublished discussion papers (Taylor, 1968; Spence, 1967), and was originally presented at the International Geographical Union Commission on Quantitative Methods Conference, London School of Economics, August 1969.

develop this analogy here as this has been adequately done elsewhere, especially in Grigg's work, so that we will only come back to it, from time to time, where necessary in describing techniques.

The full fruits of this analogy have only begun to be realised as regional taxonomy and other taxonomies have begun to use a common language—mathematics. McNaughton-Smith (1965, 1) comments:

> As taxonomy in its widest sense has moved from being a semi-intuitive art of classification towards the use of more objective methods, workers in many fields have developed numerical techniques. . . .

By way of introduction we will expand on this statement, first of all by tracing briefly the development of quantitative techniques within the broad field of taxonomy, secondly by attempting to justify this development and then finally by describing the framework in which these techniques are considered.

I Introduction

In the nineteen-thirties there emerged what became known as the 'New Systematics' (Huxley, 1940). Although this development enjoyed the now customary prefix 'new' it in no way heralded a 'quantitative taxonomy' (Sokal and Sneath, 1963, 4–5). Sokal and Sneath regard this lack of quantification as being due to the failure to separate the two scientific concepts, description and hypothesis. In fact taxonomists of this pre-quantitative school distinguish between the 'typology' approach, which is purely descriptive, and the 'biological' approach which incorporates phylogenetic hypotheses in the classification (Inger, 1958, 371; Simpson, 1951). With this latter approach taxonomists weighted morphological elements according to their suspected importance in evolutionary terms. Thus the 'experience' of the taxonomist is emphasised and the subject is categorised an art (Simpson, 1961). Such a stage of development is obviously not conducive to the growth of quantitative techniques, so that advances in biometrics such as Fisher's (1936) development of discriminant functions had little impact on the subject.

It is of interest to note that Grigg (1965) in his regionalisation–classification analogy finds a parallel between the phylogenetic basis of biological taxonomy and the geographical deterministic basis of much regional geography as discussed by Wrigley (1965). Indeed regional geography was traditionally considered an art and has been criticised as such (Kimble, 1951). No such situation inhibited the use of quantitative methods to solve regionalisation and classification problems in all disciplines. In fact American sociologists were incorporating location variables (Hagood *et al.*, 1941) and contiguity constraints (Hagood, 1943) in regionalisation algorithms a quarter of a century before the same approaches were applied by geographers. At the time of Huxley's 'New Systematics' clustering procedures were being developed in behavioural taxonomy (Holzinger, 1937, chapter 3; Cattell, 1944) following on from the earlier development of factor analysis—a detailed history of which can be found in Cattell (1965)—and of Tryon's 'cluster analysis' (1939). Similarly Fisher's discriminant analysis became widely used in statistics and was applied to the problem of optimal stratification in sampling (Cochran, 1963). Quantitative techniques also have long been used for classification in ecology (Sorenson, 1948; Goodall, 1953; Greig-Smith, 1964). In the taxonomy of the animals and plants themselves such developments had to wait for the liberation of the classification process from phylogenetic considerations (Sokal, 1965). This occurred in the late nineteen-fifties with the 'numerical taxonomy' school of Sokal and Sneath, which again parallels the application of quantitative techniques to regional problems (Zobler, 1957, 1958a, 1958b; Berry, 1961a) as Grigg (1965) has discerned. We will consider this dual development in a little more detail.

Berry (1966, 1967a) uses the term 'numerical taxonomy' to describe the regionalising procedure he has pioneered. Sokal and Sneath (1963, 37) define numerical taxonomy as 'the numerical evaluation of the affinity or similarity between taxonomic units and the ordering of these units into taxa on the basis of their affinities.' However it is necessary to point out that they go on to say that 'it is based on the ideas of the eighteenth century biologist Adanson.' Their subsequent 'six basic positions' or axioms are somewhat restrictive and by this definition Berry's procedure would not qualify. For instance, one of the axioms is the equal weighting of all characters. These ideas are not universally held among quantitative taxonomists (for example McNaughton-Smith, 1965, 11) and Sokal and Sneath's concept of numerical taxonomy may well perhaps be best described as 'Adansonian taxonomy' (see for example Wayne, 1967). This has become one of several approaches within the broad field that has been termed 'taxometrics'. These approaches will be discussed in subsequent sections, but before we come to consider the quantitative methods themselves we will attempt to justify their presence in procedures that have long been carried out in their absence.

Sokal and Sneath (1963, 49) identify 'two outstanding aims' in their work namely 'repeatability' and 'objectivity':

> Although we cannot expect scientists always to agree on interpretation of facts, it is the aim of scientific methodology to reach agreement on the facts themselves through the repeatability of observations.

Thus

Classification must be freed from the inevitable individual biases of the conventional practitioner of taxonomy.

This sentiment is echoed in regional taxonomy by Bunge (1966b, 376) who claims that 'at long last we can replicate our regions.' Johnston (1968, 576) takes exception to this statement in discussing 'the subjectivity of objective methods', which brings us to the second of Sokal and Sneath's two 'aims'.

The question that has been asked is simply: How objective is taxometrics? For instance Inger (1958, 371) writes:

It is a common assumption that operations involving numbers are more rigorous or more objective than those involving words. In actual practice this is not necessarily true.

More generally Chisholm (1964, 94) has argued that all classification is subjective:

Notwithstanding that much quantitative material is used in building up a classification, the inherent nature of the problem means that a classification must remain subjective.

This is in fact arguing the concept of objectivity out of existence. As Sokal and Sneath note, 'it would hardly seem necessary to stress that, like most rules of scientific methodology, objectivity is a relative concept, seldom fully realised.' Quite obviously 'No perfectly rational classification of phenomena exists independent of the use to which the classes are to be put' (Warntz, 1968, 10). The point is that given some research purpose, or more specifically an objective function, we require procedures to produce the one solution that is most useful for the research purpose, that is, the optimal solution in terms of the objective function. In such a situation we could quite reasonably claim repeatability and objectivity. Theoretically this situation is possible: we can arrange the base units into every possible solution and then choose that which best satisfies our objective function (Thorndike, 1953). However, Friedman and Rubin (1967, 1164) point out that:

The computational problems of evaluating all partitions of n objects into g groups in order to select one that maximises a given criterion function is solvable in principle but not in practice since the number of partitions is enormously large.

This same point is made by Hartigan (1967), King (1967) and Jones (1968), while both Scott (1969a) and Edwards and Cavalli-Sforza (1963) give examples. One problem reported by Scott, simply involving the optimum partition of 25 points in a two-dimensional space, required 50 minutes to achieve a solution on an IBM Series 360 computer, while Edwards and Cavalli-Sforza mention that in their quite exhaustive procedure 'up to 16 points may be treated in a reasonable time.' The above line of reasoning is responsible in part for the historical development of some of the characteristics of objective functions used in taxonomy. The early ones were concerned mainly with the similarity and links between the observations to be classified. Many required only hand computation and the more sophisticated ones only relatively simple computer programs. Then with the increasing use and availability of large computers, more comprehensive objective functions were developed in which the basic internal structure of groups could be tested. However even now, as is argued above, in many taxonomic problems we must usually be satisfied with a sub-optimal solution. Innumerable sub-optimal procedures have been suggested but unfortunately the form the solution will take has not always been explicitly specified. The result is that the taxonomist is unsure as to which technique best suits his problem, so that two research workers with the same problem may choose different techniques and therefore produce different solutions. Thus we are not replicating our classification as Johnston (1968) points out. It is to be hoped that this situation merely represents an early phase in the development of taxometrics. In this situation there exists a need for comparative studies between techniques such as those by Mannetje (1967), Lankford (1968) and Johnston (1968), and in particular the studies of Sokal and Rohlf (1962) and Hartigan (1967) who consider ways of comparing hierarchic solutions. Ideally we should be working towards the situation where a set of taxonomic problems are identified and the appropriate optimal or sub-optimal procedures are specified for each problem. Thus given the same problem, taxonomists will select the appropriate procedure and each produce the same solution. Here replication is present and a high degree of objectivity can be claimed.

Bearing the above points in mind, the framework within which methods are described below consists of three parts. In section II procedures are described whereby solutions are produced agglomeratively, that is, the classification or regionalisation is formed by combining base units. The alternative approach, the division of the total population into classes, is covered by the techniques described in section III. These two types of procedure have, of course, long been recognised as the basic approaches to classification and regionalisation and, in fact, form part of Grigg's (1965) analogy. The two approaches are referred to in the taxonomic literature (Simpson, 1961) and regional geography literature (Hartshorne, 1939; Whittlesey, 1954; Gilbert, 1960) where the terms synthetic regionalisation and analytic regionalisation have come to be used (Grigg, 1965). Finally in section IV we consider the problem, not of forming regions, but of testing given regional structures and the allocation of new units to them. In these three sections we describe a wide range of methods and have tried to place them so that techniques producing like solutions tend to be considered together as a group. Thus it is hoped that the following consists of more than a simple listing of techniques and at least indicates to the research worker the group of procedures in which he can find the appropriate technique for his particular regionalisation problem.

II Agglomerative procedures

Although the principal subject matter of this section concerns grouping strategies, a discussion of some of the preliminary operations should be briefly outlined. Such operations common to all taxonomic methods *per se* can if required be divided into those involving some form of *ordination* and those used to derive measures of *similarity*. In this way a logical structure should emerge indicating how a set of variables over a group of observations can be ordinated, scaled according to some measure of similarity and finally grouped or divided. This corresponds to the three stages of numerical regional taxonomy outlined by Berry (1967a, 75). However, as will be illustrated later the process need only be two-stage, the ordination procedure being omitted. This latter approach is characteristic of much of the work of Sokal and Sneath (1963).

1 Methods of ordination

The ordination stage can be theoretically envisaged as a summary of the information contained within the observation/variable data matrix. This summarisation process is attained by transforming the original geometrical space defined by the observations and the variables into another space having fewer or more basic dimensions. The process is relatively simple in theory, but perhaps not so simple when the methods involved are examined. There are several methods of ordination which can be employed in regional taxonomy, and which can of course be used for numerous other purposes. They will not be discussed in great detail here because they form a major focus of study in themselves and they are of relevance to the present subject only in broad notional terms. The ecological literature provides the best source of reference in this field (Orlocci, 1966; Austin and Orlocci, 1966; Greig-Smith, 1964). Indeed, one of the more interesting uses of these procedures in a geographic context is provided by non-geographers (see Holloway and Jardine's [1968] study of zoogeographic regions).

The process of ordination can combine the aims of scientific parsimony with that of the attainment of some pre-required special property of the data set. The methods range from elementary ordinations using a variety of distance measures and different methods of constructing simple reference axes, to the wide variety of progressively more complex multidimensional scaling methods (Bassett and Downs, 1968). Multidimensional scaling methods are used to analyse what Bassett and Downs, after Cattell (1966b), term the two-dimensional facets of the data box. In essence this is akin to the notion of the generality of the geographical data matrix propounded by Berry (1964). Included within the range of these methods are a wide variety of factor analytic procedures, more respectably termed empirical eigenvector methods. Another broad range of ordination methods are concerned with non-metric multidimensional scaling. These have been developed for the most part in response to the limiting assumptions of the factor analytic model based on its characterisation as a linear model Euclidian space (Kruskal, 1964a; 1964b; Shepard, 1962).

As all methods of ordination use similar theoretical principles, attention will be focussed on the use of principal components/factor analysis as they have been most widely used in regional taxonomy. These are at present the most efficient methods of ordination and have been of value in regional taxonomy in several ways since Kendall (1939) and Hagood *et al.* (1941) first used them. Some of the following points are more general and have been raised by Rummel (1967). There exist several bibliographies and commentaries on these methods within the geographical literature (Thompson and Hall, 1969; Thompson, 1967; Broadbent, 1968; Spence, 1967).

Factor analytic methods can be used to disentangle unknown interdependencies in a particular set of data. If the taxonomist intuitively reasons that a particular data matrix has several complex sets of interrelated variables, factor analysis will objectively isolate the interrelated data. Goddard's study of the economic activity structure of the City of London (1968) is a good example of this notion. A principal components analysis was undertaken on 80 activity types of employment recorded over some 216 street blocks. Theoretically one would expect certain groups of city centre activities to be linked both in business and spatial terms. This is exactly what the principal components solution produced. Factor one showed a trading orientation made up of the highly linked commodity trading, risk insurance and shipping activities. What was termed the 'financial ring factor' made up the second dimension and was principally concerned with capital and investment finance. Factor three could be seen as a publishing and professional services factor which is highly linked and localised. The fourth factor was determined by textile trading and other manufacturing variables. Finally, the fifth dimension isolated the other side of the financial activities of the city, namely the day-to-day working of the banks and the money market. In such a way a complicated and interdependent set of data can be transformed into a set of clearly defined clusters of interrelated data on which subsequent classification can be made. Other examples of this use include Berry (1960, 1961b), Hattori *et al.* (1960), Henshall (1966) and Henshall and King (1966).

Scientific parsimony is another objective of factor analytic methods. A study of urban dimensions of the United States in 1960 illustrates this point (Berry, N.D.). This study is based on a factor analysis of 1,762 urban places with populations exceeding 10,000 people over a series of some 97 primary socio-economic variables. 14 independent dimensions of variation were produced on which a subsequent classificatory study was based. Out of this extremely complex data set the first factor represented the functional size of centres in the urban hierarchy, reflecting such variables as size of population, and isolating the major national business centres. The second factor could be interpreted as that of the socio-economic status of the community residents relating to income, education and housing variables, and delimited politically independent sectors of rich metropolitan areas as the high status communities and the rural remote regions as the low status ones. Factor three was characterised by the stage in the family cycle reached by community residents. This was mainly concerned with the age structure of the population, and not surprisingly cities with predominantly young populations and those with ageing populations were isolated at the extremes. Factor four exhibited a non-white population dimension, and so on until all of the 14 factors could be discovered. The important point to note here is that a massive set of 97 primary variables could be reduced to 14 factors with a loss of only 23% of the original variance in the data set. Many other principally taxonomic studies can be cited to illustrate this point, for example Spence (1968) used a principal components solution to reduce a matrix containing some 60 employment variables recorded over British counties to a basic 8 factors which in total accounted for over 70% of the original variance. Similarly in a study of the urban structure of London, Dear (1969) managed to reduce his original 80 observations by 40 variables data matrix to an 80 × 5 factor matrix losing only about 10% of the original variation in the data. This made the subsequent cluster analysis of the areas of the city more manageable. Other workers have provided examples of this use, for example Megee (1965), Prakasa Rao (1953) and Thompson *et al.* (1962). By engaging in this form of scientific parsimony a large number of variables can be used in a taxonomic study without the associated problems—interpretative as well as computational.

Another use of factor analytic methods prior to taxonomic procedures *per se* concerns the exploration of the basic structure of a data set. One can of course see some relationship to previous discussion of interdependency, but these discussions of the uses of factor analysis in taxonomy are not meant to be mutally exclusive. The legion of taxonomic factorial-ecology or social-area analytic studies shows this use to be important. In recent work by Berry and his students a three-fold approach to social area analysis has emerged. The traditional approach is to derive the three basic indices of social/economic rank, family status and racial status (Bell, 1953; Shevky and Bell, 1955). This is termed social area analysis *sensu stricto*. A variation of this is a simple factor analysis of *sensu stricto* variables such as the study by van Arsdol *et al.* (1958). The third and new approach (see for example Murdie, 1968) is that of factorial ecology (Rees, 1969). This is basically a factor analysis of a wide range of variables from which the basic urban structure is sought. The studies in factorial ecology have isolated three similar sets of variables—the socio-economic status set; stage in life-cycle or family status set; and finally a minority group set (Berry and Rees, 1969). However all three of these sets need not occur in every urban structure, nor need they be mutually exclusive. As a result several alternatives can occur. First there can be a direct correspondence of the above variable data sets with the resulting first 3 factors. This is the true Shevky–Bell situation. Second the economic and family status variable sets can combine to make up the first factor, as was found in a study of Cairo (Abu-Lughod, N.D.). Third the economic and racial status variable sets can combine to form factor one. This characterises the urban structure of cities in the American south (Cohen, 1968; Peters, 1968; Spodek, 1968), Montreal (Cliffe-Phillips *et al.*, 1968) and Helsinki (Sweetser, 1965a, 1965b). A fourth combination is that of the family and racial status variables to form a second factor; this is confirmed by a study of the factorial ecology of Miami (Caswell, 1968). Chicago illustrates yet another alternative where the economic and racial status structures are linked although not fully combined (Rees, in press). There are many other alternatives which could be cited, coming mainly from the centre for Urban Studies at the University of Chicago, but the main point about the use of factor analysis in defining structure on which social area typologies can be derived has been made.

Although, as Rummell has pointed out (1967), factor analytic techniques have a variety of other uses such as exploration, systematic mapping, hypothesis testing and theory development, they are nearly all of marginal interest to the taxonomist in the sense that they really help only to understand the original data set to be classified. However there is another use of factor analysis which is important to the regional taxonomist and this corresponds most nearly to the general concept of ordination in ecology. This use of the technique is simply a data processing operation which transforms and scales the particular data set. Often the taxonomist requires a scale on which variables, such as diseases (Pyle, 1968), voting behaviour (Russett, 1966, 1967), rural poverty (Berry, 1965), space preference (Gould, 1967a; Peterson, 1967) or housing patterns (Carey, 1966) can be compared and ranked. Such a scale is achieved in this type of multivariate analysis through each dimension representing an independent or orthogonal axis of variation on which an individual variable or observation can be rated. This type of data transformation has important consequences for regional taxonomy, as well as for other types of analysis such as multiple regression. Such consequences are concerned with the frequent requirement of taxonomic procedures, especially those using Euclidian distance, for an orthogonal observation/variable space. Factor analysis transforms a highly interrelated and interdependent data set into a series of basic orthogonal or independent dimensions. A detailed mathematical description of factor analytic techniques is not presented here, but the reader is referred to Berry (1967a), L. J. King (1969) and Gould (1967b) in the geographical literature, and to Cattell (1965, 1966a), Harman (1961), Lawley and Maxwell (1963) and Cooley and Lohnes (1962) in the statistical literature for comprehensive accounts. Two interesting and possibly important developments in this field are, first, the introduction of 'three mode factor analysis' in which a three-dimensional matrix, perhaps incorporating a temporal as well as a spatial dimension, can be analysed (Tucker, 1963, 1964). This provides a possibility for dynamic ordination on data for which previously a comparative static approach has been used (King, L. J., 1966; Spence, 1969b). Secondly there is the recent use of a bifactor analytic approach (Jeffrey *et al.*, N.D.) again to study the fluctuations in an economic time-series recorded over a spatial series.

It is perhaps surprising in view of the context which factor analytic techniques have been given in this paper that Rummel has listed as a use of factor analysis, classification itself. This is a controversial point, especially in the ecological literature (Pears, 1968; Goodall, 1954; Kershaw, 1964). In this field two schools of thought exist—those who believe classification in the true sense to be impossible because vegetation is continuous and devoid of any naturally occurring discrete units; and those who do not. If the first notion is accepted, methods of ordination, such as factor analysis, are appropriate by themselves. However if the notion of classification is accepted and required, then the ordination procedures by themselves will give only a broad indication of classes. An ordination procedure has been used in this way by Tryon (1955) who locates census tracts in a three-dimensional space so that he can visually identify classes. A similar example which maps factor scores before visually grouping is the study of the socio-economic demographic characteristics of British towns by Moser and Scott (1961).

Such procedures are extremely subjective and can utilise only a three-dimensional factor space at the most. As a result the more important developments in regional taxonomy require the development of further generalised multidimensional measures of similarity as preliminary to grouping and divisive procedures.

2 Measures of similarity

Following Berry most agglomerative regionalising has involved the three-step process of which this section is a part. However this is by no means typical of all taxonomy, a simpler two-stage procedure being very common. This misses out the ordination stage altogether and begins by measuring directly the similarity between the observations or *operational taxonomic units* (OTU). This approach is put forward by Sokal and Sneath (1963) who recognise three types of measures of similarity—coefficients of association, correlation coefficients and distance measures. A fourth type of measure has been added, the probability approach, which has largely been developed since 1963. Most of these measures are found in general use outside the field of taxonomy and so will not be described in detail. However we will briefly consider each type in turn referring to the major references in the field.

a Coefficients of association: Coefficients of association have a long history of development, especially in ecology (Cole, 1949, 1957). They involve presence/absence data arranged into a 2×2 frequency table of OTU against OTU. Obviously the higher the frequency in cell 1,1 (where both OTUs possess the same attributes) and cell 0,0 (where both OTUs lack attributes), the greater the resemblance between the two OTUs. Thus all the coefficients of association consist of ratios between some combination of 1,1 and 0,0 cell frequencies and a measure of the total number of frequencies. The most elementary is Sokal and Michener's (1958) *simple matching* which is the ratio of the sum of the frequencies in the 1,1 and 0,0 cells over the total frequency of all four cells. This is just one of 16 measures reviewed by Sokal and Sneath (1963, 125–41) which differ from one another in terms of frequencies counted for the denominator and the total frequencies used as numerator. Sokal and Sneath (1963, Table 6.1, 129, and Table 6.2, 132) have used this variation to propose a useful typology of the coefficients, and the reader is referred to it for further consideration of these measures.

Of course, the best known measure of association is the χ^2 statistic. This has been widely used in taxonomy particularly in divisive approaches and so will be discussed in section III. This measure can, of course, be used with larger contingency tables and can also be converted into a correlation measure (the contingency coefficient) which belongs to the second type of measure of similarity.

b Coefficients of correlation: When dealing with data other than on a binary scale some other measure of similarity must generally be used. Correlation measures exist for all levels of measurement. The non-parametric measures—the *contingency coefficient* for nominal data, and *Spearman's rank correlation* and *Kendall's rank correlation* coefficients for ordinal data—are all described in detail in Siegel (1956). The most commonly used correlation measure of similarity is the *Pearson product moment correlation coefficient* described in most statistical textbooks. Examples of its use in taxonomy are cited by Sokal and Sneath (1963, 141–2). However it should be pointed out that not all taxonomists believe this coefficient to be a suitable measure of similarity (see, especially, Eades (1965)).

c Probabilistic measures: Although many of the measures above derive from statistics the approach is not directly probabilistic. As Goodall (1964) points out they do not specifically define the probability of whether two OTUs belong to the same population or not. His *probabilistic similarity index* is just such a measure, and another, closely related, is the *deviant index* (Goodall, 1966) which measures the extent to which an OTU differs from the 'norm' of a population. This 'norm' can be defined as an average measure, median or mode, depending on the level of measurement, and the index simply involves calculating the proportion of other OTUs farther from the norm than the OTU under consideration.

As well as these measures developed by Goodall, taxonomists have also begun to use indices defined in terms of 'information'. These are discussed by McNaughton-Smith (1965) and have been used, for example, in the Hyvarinen (1962) clustering procedure described below. Recent work in this field can be found in the papers by Wallace and Boulton (1968) and Lance and Williams (1968b).

d Distance measures: Given the co-ordinates of any two points on a plane the distance between them can be easily calculated by using the Pythagoras sum of squares equation. This equation can be generalised to measure distances between points in n-dimensional space, given co-ordinates. Thus:

$$d_{ij} = \sqrt{\sum_{z=1}^{n} (x_{iz} - x_{jz})^2}$$

where d_{ij} is the distance between points i and j in a space of n dimensions with x_{iz} and x_{jz} being measures along the zth co-ordinate. This is the simplest distance measure and is used by Berry (1961a). Several other measures are available such as Pearson's (1926) *coefficient of racial likeness* and Mahalanobis's (1936) *generalised distance* (Rao, 1948). These and other distance measures are described in Sokal and Sneath (1963, 143–53). Basic references are Sokal (1961) in a taxonomic context and more generally Blumenthal (1953).

Although there have been several comparative studies of these various measures of similarity (for example, Sneath, 1961) any discussion of the relative usefulness of various measures remains tentative. More recent comparative studies have varied similarity measures along with grouping strategies (Mannetje, 1967), and it would seem that specifying a similarity measure will become an important aspect of defining an objective function. Thus detailed investigation of the properties of each measure is required so that the research worker will be able to decide which measure is most relevant to his specific research purpose.

In the absence of this knowledge it may be noted that the Euclidean distance measure defined above is conceptually the most straightforward. This measure would seem to be most suited to the three-stage agglomerative procedure of Berry, although it has been used in a two-step procedure by Stone (1960). Stone calculates distances between 12 United Kingdom Civil Defence regions in an 11-dimension space based on 11 regional accounting variables. It is important to note however that the Euclidean distance measure should only be used where the space is orthogonal. Where axes are not at right angles we have moved into affine geometry (Sawyer, 1966, 29) where many of the properties of Euclidean geometry no longer apply, including Pythagoras' theorem. This problem does not occur if an orthogonal ordination stage precedes the distance measurement. Thus after this type of ordination, distances between OTUs in the space can be calculated and arranged to form a distance similarity matrix. When only a two-stage procedure, as described by Sokal and Sneath is used, the similarity matrix is produced in just one step as the initial stage. However, in all cases such a similarity matrix forms the starting point of the final stage in agglomerative procedures—the grouping.

3 Grouping strategies

In the use of all but one of the techniques that come under this heading we assume that some *connection matrix* is given. This may be a *similarity matrix* of distances or correlations between OTUs obtained by the methods outlined above or else, if we are concerned with functional regionalisation, the matrix cells will contain information or inter-area flows. An interesting example of an interaction measure is Ng's (1968) *W index* which measures the proportional amount of mutual movement between two areas. Other examples in geography can be found in the work of Goddard

(1970), Garrison and Marble (1963) and Nystuen and Dacey (1961), while similar interaction data has been used in other disciplines (Cattell *et al.*, 1966; Festinger, Schachter and Back, 1955). Thus the connection matrix may display connections in terms of interaction or similarity. However most grouping strategies have been proposed with similarity connection in mind and therefore our description of them will most often be in these terms.

For the requirements of the ensuing discussion we define two terms. A *threshold* is a critical value at any single stage in the grouping which denotes the level of connection necessary for potential group entrance. A *link* is any one of these connections above the threshold. Three basic approaches to grouping can be recognised:

1 *The matrix extraction* approach involves some manipulation of the connection matrix usually followed by some simple process for extracting groups. Although they produce a solution at a single level of generalisation, the number of classes is not chosen and so 'natural groupings' may be claimed. However, these procedures do usually involve the choice of a single threshold.
2 *Hierarchial groupings* usually involve a varying threshold. OTUs that are highly connected are grouped together at an early stage, and as the threshold is lowered more OTUs or groups merge to form classes at different levels of generalisation until all OTUs are joined to form a single class. Thus the solution is a hierarchy of classes and is typically illustrated in the form of a tree diagram or dendrogram.
3 *Clustering procedures* may involve a varying threshold but they do not produce classes at different levels of generalisation. However, unlike the simple matrix extraction, the level of generalisation of the solution is usually specified beforehand in terms of number of classes. These procedures typically involve two stages: first identifying nuclei or cores and then allocation of other OTUs to these cores. Of course both the hierarchic system and the 'core margin' pattern have their antecedents in traditional regional literature (Grigg, 1965, 1967).

Grouping strategies can also be classified in terms of group entrance requirements. For instance Sokal and Sneath (1963) divide the hierarchic procedures they review into single, average and complete linkage which give varying degrees of compactness. However these three categories are less meaningful today, for with the application of large computers to grouping problems total comparison on an iterative basis is common (for example, Ball, 1965; Jancey, 1966; Scott, 1969a), so that group structure is controlled by specifying the appropriate objective function in terms of group variance. However in many older techniques internal structure is an important consideration, as will be seen in subsequent sections.

a Simple matrix manipulation and extraction: Before we consider the explicit extraction procedures we shall describe some simple matrix manipulations that have been used in taxonomic problems.

Conceptually the simplest grouping procedure is probably the *matrix diagonal method* (Cattell, 1944, 172; Sokal and Sneath, 1963, 177–8). This is a graphical procedure involving the shading of similarity matrix cells. If the more similar connections are shaded darker, the procedure is to rearrange the OTU ordering along the sides of the matrix so as to transfer darker shading towards the matrix diagonal. In this way any grouping within the data is graphically illustrated, but the method is both cumbersome and not practicable with large matrices. It has been used with social interaction matrices by Forsyth and Katz (1946), but has been replaced in this field by use of conventional matrix manipulations as proposed by Festinger, Schachter and Back (1955). Having produced a connection matrix describing relationships between people within housing projects they square and cube the matrix and in this way identify 'cliques' and varying degrees of 'cliquishness'. Recently another matrix manipulation process has been proposed with similar objectives. This is Bonner's Clustering Program I which uses a method of converting an attribute/OTU matrix into an OTU/OTU similarity matrix, but Bonner applies it to the latter so that he is 'taking the similarity matrix of a similarity matrix'. The procedure can be iterated as many times as required taking one minute per iteration with a 350 × 350 matrix. The result is 'to give a better definition to clusters which are loosely connected internally and to better separate those that overlap' (Bonner, 1964, 23).

These three procedures all have the same limitation. Although they simplify the connection matrix they do not explicitly define classes, that is they do not include specific procedures for abstracting groups from the matrix. It is to such procedures that we now turn.

In the matrix manipulation of Bonner the connection matrix has to be converted into a binary matrix by using a single threshold and describing all connections above this level as unity and all those below as zero. This preliminary transformation is also required in the following abstraction procedures. Using just such a binary connection matrix the same definition of groups has been made independently in psychology by Cattell (1944), in palaeozoology by Olson and Miller (1958) and has been used in medical research by Bonner (1964). This definition is simply that a group consists of a set of OTUs where each OTU is linked to every other member of the set and no other OTU in the population is linked to every member of the set. Such groupings are called *homostats* by Cattell (Cattell *et al.*, 1966), *tight clusters* by Bonner (1964) and *ρ groups* by Olson and Miller (1958). Their abstraction from a matrix is described by Cattell as the *ramifying linkage method,* with the *approximate delimitation method* where large matrices were involved. Cattell has incorporated this group definition as the first step in his *Taxonome* computer program (Cattell *et al.*, 1966, 314–17). Similarly this repetitive procedure is computerised by Bonner (1964) as part of his Clustering Program II and he also refers to another algorithm for finding tight clusters by Needham (1962).

These algorithms will usually produce a complex solution of overlapping structures so that in our examples above the definition of these groups is considered a first stage in the taxonomic process. However each differs in their subsequent refinements. Cattell goes on to search for what we have termed 'natural' groups and his approach is described under that heading below. Both Olson and Miller (1958) and Bonner (1964) develop their procedures so that overlaps are excluded, but not in such an extreme form as Cattell: they merely aim to produce disjoint sets of OTUs. Bonner's algorithm for producing *disjoint core clusters* comprises the second part of his Clustering Program II. It involves first of all defining the largest right cluster as the first core cluster, and then searching in its complement for the largest tight cluster to become the second core cluster. The procedure iterates on these lines and the complete algorithm is laid out in Bonner's (1964, 25) Table 3.

Olson and Miller take a different approach in deriving *basic pair ρ groups*. A basic pair is simply two OTUs which are mutually most similar. After listing all ρ groups they reject those that do not contain both members of a basic pair. OTUs are also dropped from a group if they are more similar to a member of a basic pair outside the group than with any basic pair member in the group. This procedure will tend to lead to discrete groups but will not do so in two circumstances:

1 where there is a tie in the similarity between an OTU and two basic members in two groups
2 where a basic pair occurs in an intersect between two groups.

Olson and Miller suggest a procedure for overcoming the first type of overlap but the second type remains and will mean that unlike Bonner's development these procedures will not necessarily produce discrete groups.

In these latter parts of Olson and Miller's procedures an ordinal (i.e. 'first ranked') criteria is being used, for instance in defining basic pairs. There is a whole series of techniques that are wholly ordinal and it is to these that we now turn. The simplest is Rice's (1965) method which is largely graphical. Each OTU is drawn as a point and a solid line connects it to the OTU to which it is most similar and a dashed line to the OTU to which it is second most similar. These diagrams then 'suggest' groups—Rice does not propose any definite rules as a procedure for identifying the groups. However McQuitty (1957) has suggested a procedure which produces discrete groups while also being much simpler than Olson and Miller's basic pair groups. The threshold is the ordinal 'first ranked' criteria from which a binary connection matrix may be produced as with an interval threshold. This matrix is the basis of McQuitty's (1957) 'Elementary linkage analysis'. For a geographical example see the temporal unemployment study of some 30 mid-western standard metropolitan areas by Jeffrey *et al.* (N.D.). In this procedure every OTU is allocated to the group in which the OTU to which it is most similar is found. By starting with reciprocal pairs (a mutually similar pair) a set of discrete groupings can be quickly found in which every member of a group is more similar to one other OTU in the group than to any other OTU. McQuitty shows that each group will have one reciprocal pair and that there will be no residual OTUs. Such groups are described by Haggett (1965, 283) as *basic pair correlation sets*. They differ from Olson and Miller's basic pair ρ groups in two ways. Firstly Olson and Miller's groups (and Rice's) can have several reciprocal pairs in a single grouping. Secondly Olson and Miller keep an interval level of measurement in their analysis by maintaining their original group threshold. For instance any reciprocal pair whose connection is below the interval threshold is not considered a basic pair and, along with other OTUs with no connection above the threshold, they remain isolated throughout the procedure. These differences suggest two criticisms of elementary linkage analysis. First of all entrance to a group is based on a single highest link so that chaining is a definite possibility leaving very loose groupings. Secondly the technique will produce just as many groups as there are reciprocal pairs, a property of the connection matrix which is not necessarily so relevant.

Dacey (1958) has recognised these limitations and has suggested an alternative procedure. In his 'Nearest neighbour linkage analysis' he uses the concept of *incremental distance* (I)

$$I_{ij} = | d_{il} - d_{ij} | + | d_{jl} - d_{ji} |$$

where d_{il} and d_{jl} are first ranked connections for OTUs i and j respectively and d_{ij} and d_{ji} are connections between i and j and j and i. If i and j are a reciprocal pair I_{ij} is zero. Thus incremental distance can be thought of as a measure of the deviation of a pair of OTUs from the ideal reciprocal pair. Dacey arranges McQuitty's (1957) original data into groups by allocating OTUs so as to minimise incremental distance and produces four classes as opposed to McQuitty's two. Thus the grouping is no longer dependent on the number of reciprocal pairs, and as 'the degree of reflective similarity' is maximised there will be less tendency towards loose groupings. Examples of the use of both techniques can be found in Norcliffe (1968).

McQuitty (1963) has developed a technique which produces very compact clusters. He defines the groups as sets in which each OTU is closer to *all* other OTUs in the group than to any other OTUs in the population. This is achieved by ranking all connections in the matrix and then adding OTUs to reciprocal pairs under specified conditions. The exact procedure is described by McQuitty (1963) and Johnston (1968, 583) and is known as *rank order typal analysis*.

In general these matrix extraction procedures have not been described in great detail as they can usually be considered simply the first step in a more comprehensive taxonomic analysis. This is explicit in the work of Bonner and McQuitty, the former going on to develop what we have defined as a clustering procedure and the latter several hierarchical techniques which we will consider first along with other hierarchical techniques.

b Hierarchical techniques: Having carried out one of McQuitty's linkage analyses the resulting classes can be treated as new OTUs, a fresh similarity matrix constructed and the linkage procedure repeated to produce a system of classes at a second hierarchical level. This process can continue until one class remains. A whole series of methods for accomplishing this procedure exists and they vary in the way they define new OTUs and in their similarity measures in the new connection matrices.

Similarity measures within new groups can be calculated between mean points of groups which Johnston (1968) terms centroid replacement. Alternatively the two OTUs of the two groups that are furthest apart can be used (Johnston's 'replacement assumption') or in contrast the two that are nearest can be used (Johnston's 'replacement assumption *B*'). The first two ways of measuring similarity are used by McQuitty (1966) and the third one is suggested by Johnston (1968). The OTUs of the similarity matrix can be:

1 groups formed by elementary linkage analysis
2 groups formed by rank order typal analysis with the residual OTUs
3 all reciprocal pairs plus residual OTUs
4 the most similar reciprocal pair plus the residual.

These first three ways of defining OTUs and the three ways of defining new similarities mean that there are potentially nine procedures, some of which are illustrated by Johnston (1968). As the basic procedures are merely repetitions of matrix extraction methods already described we will not consider them further here.

The fourth definition of new OTUs above is simply to use the reciprocal pair with the strongest connection. This is used in McQuitty's (1960) 'Hierarchical syndrome analysis' and is also illustrated by Johnston (1968, 582–3). Here we can identify a change in procedure. Rather than a whole hierarchical level being identified at each stage, the classification is built up with small incremental steps. The procedure is much slower, but with recomputation of the similarity matrix after every small step, misclassification is much less likely. Thus in general it has been more widely used in biology (Sokal and Sneath, 1963), psychology (Johnson, 1967) and geography (Berry, 1961a) and it is to these procedures that we now turn.

Johnson (1967) has developed two hierarchical clustering schemes which are wholly ordinal and thus very similar to variants of hierarchical syndrome analysis. Johnson combines the nearest two in his matrix and recomputes the matrix using the smallest distances to any member of a group in his *minimum method* or the member of the group furthest away in his *maximum method*. These obviously correspond to what Johnston (1968) has identified as 'replacement assumption *B*' and simply 'replacement assumption'. Both these methods are computationally very simple and are feasible with quite large initial matrices. For instance Johnson (1967) uses a Fortran program that produces a hierarchical clustering scheme from a 64 × 64 matrix for both methods in just 10·1 seconds on an IBM 7094.

An even quicker method is the *single linkage method* of Sneath (1957) described in Sokal and Sneath (1963, 180–1) which requires no recomputation of the similarity matrix in arriving at a hierarchical solution. An interval threshold is lowered in small steps and groups develop on the basis of a single link above the threshold at any particular level. Therefore two groups may merge if just one member of each group are linked above the threshold. Unlike the interval threshold procedures of Cattell, Olson and Miller, and Bonner, there is what may be termed a discrete constraint which ensures that once an OTU joins a group it is no longer considered for individual membership of any other groups, thus avoiding overlapping solutions. This constraint is implicit in biological taxonomy (Sokal and Sneath, 1963, 179), but seems to have been explicitly considered only by Scott (1969a) and Jardine and Sibson (1968). Although Sneath's method has been criticised particularly because it is frequently subject to 'chaining'—i.e. it may produce long, strung-out groups joined together by just one or two links—it should be noted that only single linkage grouping is consistent with Jardine's (1967) *et al.* 'logical structure of taxonomic hierarchies'. Sneath has recognised the chaining problem and suggests some recalculation of similarity measures, although it would seem that if more compact groups are required it is more sensible to turn to one of the many other available procedures.

The complete linkage method developed in ecology by Sorenson (1948) only allows entrance to a group if all members of the potential new group are linked above the threshold. This technique proceeds in large steps and is more closely related to the earlier methods Johnston (1968) illustrates. The procedure is described in some detail by both Greig-Smith (1964) and Sokal and Sneath (1963).

In between this single and complete linkage comes a series of procedures typified by the *average linkage methods* originally put forward by Sokal and Michener (1958) and described in some detail by Sokal and Sneath (1963, 182–5 and Appendix A3). Two types can be immediately identified. The *pair group methods* only allow the fusion of two units in any one cycle, whereas the variable group methods allow several to join as long as they adhere to threshold and group homogeneity requirements. In the original examples of Sokal and Michener (1958) and Michener and Sokal (1957) similarity in the new matrix was calculated using Spearman's (1913) sums of variables method with correlation coefficients:

$$r_{AB} = \frac{\sum r_{ab}}{\sqrt{n_a + 2\sum r_{aa}} \cdot \sqrt{n_b + 2\sum r_{bb}}}$$

for two groups A and B with their respective n_a and n_b members denoted a and b so that $\sum r_{ab}$ is the sum of inter-group correlations and $\sum r_{aa}$ and $\sum r_{bb}$ the intra-group correlation sums. In the special case of measuring similarity between a group A and a single OTU x the equation is amended to:

$$r_{Ax} = \frac{\sum r_{ax}}{\sqrt{n_a + 2\sum r_{aa}}}$$

If similarity is measured in terms of distance then the average inter-group distance can be used (Sneath, 1962).

When these similarity measures are recalculated the problem arises as to whether to give greater weighting to more recent additions. Sokal and Sneath (1963, 185) give the example of a group *AB* joined by *C* and then by *D*. Should *C* be weighted equal to *AB* when *D*'s similarity to the group *ABC* is considered? Since in biological taxonomy late joining OTUs represent separate evolutionary lines, Sokal and Michener (1958) weight each new member as equal to the sum of all old members of the group for later similarity recomputation. However, four approaches are identified within the average linkage technique—the *weighted pair group method*, the *unweighted pair group method*, the *weighted variable group method* and the *unweighted variable group method.*

In geography Berry (1961a, 1967b) has suggested a procedure which is very similar to the unweighted pair group method. In his *stepwise grouping* he combines the nearest two OTUs in his similarity matrix and these are replaced in the matrix by their centroid. Distances from the centroid to all $(n - 2)$ OTUs are calculated and a new $(n - 1) \times (n - 1)$ similarity matrix computed for the next cycle. The procedure iterates until all OTUs are combined to form a single class. Berry (1966) has since used the above procedure with interaction data in the form of commodity flows between Indian regions. Two geographic studies of central place systems in India and Nigeria respectively have used this method (Mayfield, 1967; Abiodun, 1968). Similar procedures have been used in economics by King, B. F. (1966, 1967). A variety of objective functions could be used in the above iterative framework. The centroid method has been criticised as being only useful for data which have a definite easily recognisable pattern. This is because if the OTUs are very similar then the problem condition—'chaining'—is often the result. A more efficient objective function is the minimum increment method. This joins together the pair of OTUs which make the minimum increment to the pooled within-group sum of squares (Ward, 1963; Neely, 1965). The procedure iterates like the Berry techniques to produce a complete hierarchical tree. It was originally used by Ward and Hook (1963) and more recently it has been used in geography by Spence (1968) in a classification of British counties using employment data. A very similar procedure has been put forward by Orlocci (1967) as his hierarchical *optimal agglomeration* method. A major problem of these iterative procedures in maximising a particular objective function is the large amount of computer storage and processing time needed. Both Lankford (1969) and Neely (1965) make the point that a 32K machine can handle only 220 observations. For this reason there has been some recent attempt to predefine the major dimensions of the classification. (Neely, N.D.a; Neely, N.D.b, both cited by Lankford, 1968). This is the *neighbourhood limited algorithm* approach, which also attempts to find 'natural' groups and so is considered under that section below.

A selection of the hierarchic techniques above that have been developed in biology have been related to one another in the framework of a 'general theory' by Lance and Williams (1967), who also present computer programs for them (Lance and Williams, 1966). These 'similarity analyses' all involve 'linkage' objective functions, but Wishart (1969c) has recently shown how the Ward procedure with its more comprehensive objective function can also be incorporated in this general theory. By allowing certain transformation parameters to vary Wishart has been able to incorporate six commonly used hierarchic procedures in a single algorithm and computer program (1969c). This is of course very useful for comparative studies, but it must be pointed out that even the most comprehensive hierarchic techniques, such as those of Ward and Orlocci, despite their consideration of every possible fusion at each step, are only locally optimal (Fisher, 1958; Ward, 1963; Scott, 1969a). In general this will be true of any hierarchical arrangement of classes. Thus once any level of generalisation is chosen for further analysis this can be subjected to further iterative procedures to make it converge on the optimum solution for that scale. This we will come to in section IV. Here we continue by considering our third set of grouping techniques, the clustering procedures: these are not restricted to locally optimal solutions by the external structure of their results. For comments on the advantages and disadvantages of the hierarchic structure see Grigg (1965) and Taylor (1969).

c Clustering procedures: The earliest published attempt at a quantitative regionalisation will furnish our first example of clustering procedures. This comprises of Hagood's (1943) attempt to regionalise the United States on the basis of state population and agricultural data. Her first step was 'the easy determination of regional nuclei of states which are without question so homogenous that they should be in the same region' (Hagood, 1943, 295). 10 nuclei were identified, leaving 12 states to be allocated to them in step 2. This allocation uses correlation measures between states and seems to involve allocation to the group with the most similar neighbour, but Hagood is not specific on this point. The procedure is described in Hagood and Price (1952), Isard (1960) and Haggett (1965). The main criticism of the procedure is its inherent subjectivity, but it is useful in illustrating the two main stages in clustering procedures—first of all a definition of a nucleus and then the allocation to this nucleus. Ng's (1968) *graphic solution*, which uses his *W* index referred to above (p. 15), is similar in principle to Hagood's approach.

Hagood's method of defining nuclei is particularly unsatisfactory and would seem to be easily replaceable by one of the matrix extraction procedures that produce very compact groups. Such a procedure has been put forward by Bonner (1964) in his Cluster Adjustment Program. This involves using the core clusters defined in his Clustering Program II. A minimum size of nucleus group is specified and all core clusters above this size are taken as nuclei. Other OTUs are then reallocated to these nuclei using the original similarity matrix. The program algorithm is described in Bonner's (1964, 26) Table 4. Of all clustering procedures it is probably most like Hagood's pioneer work, for both have groups of OTUs as nuclei, whereas most of the other techniques group round a single 'typical' OTU.

The earliest definition of 'typicality' in taxonomic grouping seems to be that of Rogers and Tanimoto (1960). Using a measure of association they employ the number of OTUs with which a given OTU has a positive association as a measure of its typicality. Correlation coefficients replace the coefficient of association in one application of the technique (Silvestri *et al.*, 1962) and distance measures replace it in another version of the method (Taylor, 1969). A threshold is specified and the OTU with the most links is defined as the first node and other OTUs are allocated to it in turn. A test of group homogeneity is used to decide whether an OTU joins the group. This follows the earliest clustering procedure of Holzinger (1937) who used his *coefficient of belonging* or *B coefficient* to decide whether an OTU should join a group. This technique is illustrated by Clements (1954) with an anthropological example. In both cases when the measure of homogeneity falls sharply the cluster is removed from the study and the whole procedure repeated on the residual OTUs. Rogers and Tanimoto's *nodal clustering* has been particularly criticised by Lance and Williams (1968b) who favour the procedure of Hyvarinen (1962). Hyvarinen withdraws each OTU in turn from the population and notes the subsequent 'information loss'. The OTU that produces least information loss is defined as the most typical because it must obviously resemble a large number of other OTUs. Other OTUs are allocated to such typical nuclei and the procedure produces a set of clusters.

In the geographical literature there is a clustering procedure which uses interaction connection. This is Nystuen and Dacey's (1961) *graph theory* method which they use to define functional regions in Washington State. Given a connection matrix of flow data between base areas, column totals are computed and areas ranked in order of the total flow to them. Next every area is allocated to the area to which it sends its largest outflow. If this highest outflow is to a lower order area the link is cancelled and the area itself becomes the centre of a distinct cluster or in graph-theory terms the terminal point of a sub-graph.

Finally we need to describe the spate of new clustering procedures devised in recent years and which can all be described as iterative. They all have the characteristic of using an objective function, and then comparing a large number of possible solutions and choosing the one that best satisfies the objective function as the final solution. The procedures usually lead to an approximate optimal solution as the task of comparing all possible solutions is not feasible except for very small problems. First of all however we will describe an algorithm that does provide a global solution which is then used to test an approximate algorithm with small problems.

Scott (1969a) introduces a *backtrack algorithm* for finding the global optimal solution for minimising intra-group variance. The procedure consists essentially of an ordered search over a combinatorial tree and has been developed initially for the identification of just two groups. Scott's *approximate solution algorithm*, which complements the above procedure, consists of two basic steps involving a 'forward movement' and a 'shuffling process'. Three groups are initially defined: groups A and B and a residual group. An OTU is allocated to A and another to B. The forward movement now consists of deleting each OTU from the residual in turn and adding it first to A and then to B. In both situations the intra-group variance is noted and then the OTU is returned to the residual. The A, B, residual group content is now revised so that the OTU-group fusion which yielded the lowest intra-group variance is incorporated. The three groups next undergo a shuffling process. Every OTU is alternatively allocated to A, B or the residual in turn. For each situation the intra-variance is calculated, and if a situation occurs where a variance value is found that is below the existing level, the groups A, B and the residual are suitably altered. The program moves from the forward step to the shuffling step and back again until the residual disappears and we are left with two groups A and B. This solution will be locally optimal and will tend to vary with the initial choice of two OTUs to begin groups A and B.

Using 10 different distributions each with 20 OTUs Scott computes approximate solutions and global optimal solutions using both his algorithms. For the approximate procedure all possible pairs of OTUs were used for initial allocation to A and B and Scott obtains the very encouraging result that 1,738 or 93·1 % of the 1,900 different approximate solutions were truly global optimal results. Finally Scott compares his algorithm with another alternative approximate procedure, Cooper's (1967) *elimination-alternative-correction heuristic* and shows that his own is more powerful in converging on the global optimal. Scott (1968) also presents a discussion of other combinatorial methods with a complementary bibliography (Scott, 1969b).

One final point needs to be made concerning Scott's approximate algorithm. Although specifically developed to replace the global optimal method for large problems, its own computer time with large problems is not known. 20 OTUs in a two-dimensional space grouped into just two classes take only 2·4 seconds on an IBM 360. However both timing and standard of performance are not known for large complex problems.

An algorithm that has been used for more than two classes in several dimensions is described in Jancey's (1966) 'multidimensional group analysis'. A predetermined number, K, of classes is chosen, and K 'class points' are located randomly in the discriminant space. OTUs are allocated to the nearest class point, which is then moved to the centre of gravity of the cluster it has attracted. OTUs are then allocated to the new class points and the whole procedure iterated until the algorithm reaches a stable approximate optimal solution. MacQueen (1966), reported in Lance and Williams (1968a), and Ball (1965), suggests more flexible procedures whereby the number of classes can change during the algorithm. MacQueen's algorithm is similar to Jancey's method except that when two clusters are closer than some predetermined value, they are merged, leaving $K - 1$ classes. Also an upper limit to group size is set so that if an OTU cannot join a class for this reason it forms a new centre thus producing $K + 1$ groups. Ball's ISODA program increases numbers of classes in a different way. If any group increases its heterogeneity above a predetermined level that class is split to produce $K + 1$ groups. Another similar algorithm has been proposed by Rubin (1967), Friedman and Rubin (1967) and by Shepherd and Wilmott (1968).

The tests of these algorithms do not seem to have been as comprehensive as on those Scott has carried out. However MacQueen has used a larger population dividing 250 OTUs with five attributes into 18 clusters. This was carried out three times starting with different class points and the results agree to within 7% (Lance and Williams, 1968a). The result is not as good as Scott's, but this may be due to its being carried out on a much larger problem.

Thus we can see how clustering procedures have developed from Hagood's early work through to the iterative algorithms. These latest developments all seem to start randomly, and it would appear that the next stage in this development is to find useful criteria for choosing, as starting points, the OTUs most likely to converge on the global optimum (Scott, 1969a; Lance and Williams 1968a, 276). This completes the discussion of the three main approaches to grouping. However two further small distinctive sets of procedures exist which require some consideration. The first is related to the clustering methods above and the second consists of a single technique with many characteristic affinities to divisive methods.

d Natural clustering: Cattell has distinguished between two 'special type' concepts, the polar type and the modal type (Cattell *et al.*, 1966, 289–90). The former involves identification of extremes in a unimodal population and the latter defines 'species' about several modes. An example of the former is Cox's (1957) procedure for optimally dividing a univariate normal distribution, but, as Jones (1968) points out, such a population might usually be considered a single homogeneous group—a Cattell modal type. However most of the techniques described above will automatically produce a classification irrespective of the underlying distribution of OTUs. Simpson (1951, 1961) has termed such classification 'arbitrary' in contrast to the 'non-abritrary' or natural grouping where classes are defined by 'gaps in resemblance'. Objective functions minimising intra-group variance will not necessarily detect 'natural' clusters and this is because such clusters need not be spherical in shape. A dense elongated cluster, although clearly distinct from surrounding OTUs, will not be particularly homogeneous on intra-group variance criteria. Jones and Jackson (1967) see this problem as a choice between 'internal-coherence' and 'external-separateness'. As the latter concept is perhaps what many taxonomists are intuitively seeking, several approaches for detecting natural clusters have been recently suggested.

An example in the geographical literature is the method of Neely (Lankford, 1969) who has reverted to a fairly subjective procedure similar in many respects to Hagood's original nuclear clustering. In the neighbourhood-limited algorithm already referred to, the 'neighbourhoods' of OTUs, within which comparisons are made, are specified beforehand and these can be 'natural' in the sense above. Although this procedure may prove successful in tests carried out by a single user, the algorithm would seem to be contrary to the general objective of replication discussed in the introduction.

In psychology Cattell has recognised the need for natural clusters which he terms 'segregates' or 'aits' for short. After identifying overlapping 'homostats' in his *Taxome* program he uses the *segregate search algorithm* to find aits (Cattell *et al.*, 1966, 320–1). However a much simpler procedure has been recently suggested. The single linkage methods described above do not tend to produce spherical groups. In fact they are criticised for allowing 'chaining'. Wishart (1969a) has noticed that if OTUs in the very low density space between 'natural' clusters are removed, simple application of Sneath's (1957) single linkage clustering will produce natural grouping. This *modal nearest neighbour* algorithm is available as a Fortran II program (Wishart, 1968) and an application of it using disease symptoms (Wishart, 1969b) shows it to be far superior to the Ward algorithm and others in a situation where natural groupings were necessary. An earlier version of this procedure has been applied to areal units in a recent geographical application (Pocock and Wishart, 1969). This relatively simple procedure of Wishart's (1969a) would seem to be the best so far available for defining groups in the 'non-arbitrary' sense of Simpson. A similar but more complex type of modal analysis has been proposed by Jones (1968).

e Monothetic grouping: The technique that has been suggested by Lockhart and Hartman (1963) differs from all others in this section in that it does not start with a connection matrix. Using simple presence/absence data they first of all find the pair of OTUs that have most similarities in terms of the total set of attributes. From this point the procedures become distinctive. Similarities not common to both OTUs in this reciprocal pair are discarded and the OTU which has most similarities with the first pair is added to the group. A further set of attributes are now discarded and the process repeated. This whole procedure is carried out using several similar pairs of OTUs and the groups arranged in a hierarchy. With over 100 OTUs the above algorithm is modified so that original nuclei are randomly selected to save computer time. All groupings in this procedure have the property that at any one level of generalisation all members of a group will have a given set of attributes. This is a basic property of all monothetic classification and we will discuss this more fully when divisive procedures are considered in the next section.

4 Grouping and regionalisation

The first condition in Hagood's procedure was that 'regions shall be geographically contiguous' (Hagood, 1943, 293). It is this requirement which Grigg (1965, 1967) would argue makes the regionalisation–classification comparison an analogy rather than an isomorphism. In technique terms this condition means that when Hagood made decisions concerning regional nuclei or the later allocation, OTUs were only compared if they were locationally contiguous. In this sense the grouping procedure can be said to include a *contiguity constraint* and such classification comes under Fisher's (1958) general category 'restricted grouping'. In general 'natural' clustering would seem unsuited to any constraint and will not be considered further here.

Although originally unrestricted in the sense of Fisher, Berry's stepwise grouping has subsequently had a contiguity constraint added to it (Berry, 1967a; Ray and Berry, 1966). In effect this involves adding a binary contiguity matrix and making the procedure check that the appropriate cell is positive (that is, that the OTUs are contiguous) before attempting comparison. Such an addition has also been added to Ward's iterative hierarchic procedure (Neely, 1965) and has been used by Spence (1968) to produce a contiguous regionalisation of British counties based on employment data.

Bunge (1966c) rejects the necessity for a contiguity constraint in regionalising and simply suggests the inclusion of location as a variable in the analysis. This is accomplished by using the distance between two places as a measure of their 'locational similarity'. Thus there are two similarity matrices, one for location and another for other variables. Bunge now suggests combining the matrices with appropriate weighting and then using a stepwise hierarchic procedure similar to Berry's technique. This is unsatisfactory for two reasons: first, the combining of the matrices with a weighting seems rather arbitrary and second, the inclusion of location as a variable will not necessarily produce contiguous groups. These latter criticisms also apply to the inclusion of measures along two axes defining an areal unit's location (Taylor and Spence, 1969) as originally used by Hagood *et al.*, (1941). In any case such a procedure is not directly related to 'locational variance' or compactness and so is not the same as including location as a single variable (Bunge, 1966c). None the less Berry (1966) has included longitude and latitude among the variables used in his factor analysis of Indian data and this has drawn the interesting comment from Gould (1968) that in mapping the resulting factor scores Berry is, to some extent, 'locating location'. A more basic problem had been previously identified by Gregory (1949) who points out that Hagood's results clearly reflect her choice of axes. A clustering technique has been proposed which was partly designed to overcome some of the above difficulties. This has been termed *location-based nodal clustering* (Taylor, 1969) and is a modified form of Roger's and Tanimoto's technique with location included in the definition of 'typicality'. Thus typically is defined in terms of number of OTUs within a certain distance in both the geographical *and* discriminant space. A contiguity constraint is included both in defining typicality and in the actual grouping. Unfortunately irregular base areas mean that certain modifications have to be made and this results in the procedures which are very difficult to program, the initial example being carried out manually (Taylor, 1969).

Finally before we leave grouping we should mention a set of literature closely related to regionalising. This is the work that attempts to define 'fair' legislative districts. Algorithms are available (for example, Weaver and Hess, 1963; Hess and Weaver, 1965) which produce contiguous and compact groups of districts of approximate equal population size. Although obviously designed for a slightly different purpose, the problem remains essentially taxonomic and a study of these algorithms might repay the regional taxonomist.

III Divisive procedures

In this section we are concerned with procedures that *divide* a total population. In analytic regionalisation the division of the area was usually accomplished by means of some suitably chosen isoline. This situation certainly did not lead to a replication of regions (James, 1943; Sinnhuber, 1954) and a great deal of subjectivity was involved. The suitability of an isoline or any other divisive boundary line was usually judged in terms of some *a priori* hypothesis as Grigg (1965, 474) has pointed out. The obvious examples are the series of major world regions claiming to describe phenomena ranging from soils to agriculture but which were basically climatic in terms of their construction. We have already observed in the introduction that there is a need to separate the concepts of description and hypothesis. This has been achieved with the quantitative techniques of division largely developed in plant ecology, whereby the suitability of a variable is based on what Williams and Dale (1965, 41) term its importance *a posteriori*.

In these divisive techniques the ordination and similarity measurement procedures are either the same as those described for agglomerative techniques in the last section or else are intimately incorporated as part of the technique. Thus in this section we can go straight on to consider the divisive techniques themselves. A broader than usual view of divisive techniques is taken in this paper so that as well as the developments in plant ecology we also consider procedures from outside biology. The biological procedures are largely hierarchic and it is to these that we turn to first.

1 Hierarchic division

In the biological literature a distinction is made between *monothetic* and *polythetic* division (Sneath, 1962). This is by no means synonomous with the univariate–multivariate dichotomy of many numerical procedures. In a monothetic procedure each division is based on one variable whereas in a polythetic procedure it is based on several. The distinction is thus in terms of the number of variables used at each stage in the classification. Therefore with several stages a monothetic procedure is multivariate and it is with these techniques that we will begin.

a Monothetic division: Monothetic techniques all involve the choosing of the most 'suitable' variable with which to divide the population at any stage. Perhaps their most critical element is in the specific definition of 'suitability'.

The most commonly used method in ecology is *association analysis*. Although not known under this name until 1958 this technique was originated by Goodall in 1953. He put forward four procedures and suggested his 'Procedure I' as the best. Since this date Williams and Lambert (1959, 1960) have put forward an improved procedure which we will concentrate on here.

In all these procedures the OTUs are quadrats and the variables are simply presence or absence of plant species. Goodall (1953, 43) points out that if we are dealing with a continuous variable then the technique can be applied by dichotomising the variable about some arbitrary level such as the mean or median. In describing the techniques here the ecologist's quadrats are replaced by the geographer's areal base units and presence–absence by simply positive and negative scores which may represent a variable divided about some central location.

All methods of association analysis involve deciding whether the population of base areas is heterogenous as a first step. This is determined simply by using χ^2 as a measure of association. Every pair of variables is compared by constructing 2×2 contingency tables (positive/negative categories of each variable making up the axes) and computing the χ^2 statistic. However, this statistic is a derivative of the normal probability formula and so applies strictly only to continuous variables (Langley, 1968, 270). Inaccuracies involved in using the usual formula are small for reasonably large contingency tables but are important with 2×2 tables (Langley, 1968, 285). This problem can be overcome by applying Yates's correction so that the formula becomes:

$$\chi^2 = \sum \frac{(|O - E| - 0{\cdot}5)^2}{E}$$

With 2×2 tables this formula can be modified (see Snedecor and Cochran, 1967, 211–13; Siegel, 1956, 101). The χ^2 values are now used as measures of association between the variables making up the contingency table (but see Williams and Lambert, 1959, 86). A large χ^2 value indicates a high degree of association between the distributions of the two variables, and that the distributions are clustered among a few base areas, in other words that the population of base areas is heterogeneous. To define heterogeneity in a given problem, a threshold, in the form of a probability level, has to be chosen. Then if any χ^2 value is above the probability level the population is assumed heterogeneous and therefore in need of division.

The next step is to choose the 'indicator' variable upon which to base the division. This is achieved by arranging χ^2 values in an $m \times m$ association matrix with variables along each axis. Only significant associations are considered so that all non-significant χ^2 values are recorded as zeros. Column totals $\sum \chi^2$ now give a measure of the amount of association of the column variable and so are used as criteria for selecting the indicator variable. If the variable with the highest $\sum \chi^2$ is A then the total population is divided into positive scoring areas A, and negative scoring areas a. Both of these resulting classes are now treated as new populations and the above procedure repeated. Thus class A may be divided on variable X to produce two new classes AX and Ax, while class a is divided on variable Y and produced aY and ay. The procedure continues until no significant associations remain.

Unlike Goodall's original association analysis this revised procedure produces a simple hierarchical tree with all divisive decisions recorded in the final system of classes. Williams and Lambert (1959) conclude that their procedure is an improvement on original association analysis both for theoretical reasons and on the grounds of the empirical evidence they produce. Recently Lance and Williams (1965) have described a computer program for this technique.

An alternative monethetic procedure has been suggested by Crawford and Wishart (1967). Their '*group analysis*' is an attempt to develop a rapid method of quantitative ecological division for situations where the number of quadrats and species is too large to be dealt with by association analysis. Thus they differ in one important aspect in their procedure: they do not attempt to compare all possible pairs of species in contingency tables, but instead derive for each species a single index that describes its suitability as an indicator variable. This feature of the technique reduces computational load, but unfortunately the index does not seem to be conducive to general conversion for use with dichotomised variables describing areal units. For instance it incorporates, in part, the probability of occurrence of a species. If we translate this into use with a dichotomised variable describing a base area, the way in which it is dichotomised takes on a special importance. If the median is chosen to divide a variable, the probability of occurrence will be the same (0·5) for every variable. The same is true if the arithmetic mean is used and the variable has a symmetrical frequency distribution. Thus the procedure appears to be suitable only for use with simple presence–absence data within a region.

Despite the apparent complexity of this procedure, involving the introduction of several new statistics, Crawford and Wishart's subsequent analysis shows that it succeeds in its main objective, to produce rapid results from large sets of data. For one analysis on an IBM 1620, group analysis took only 1% of the time required for association analysis using the same data. Whereas association analysis's computational time increases linearly with the number of areas and in proportion to the square of the number of variables (Williams and Lambert, 1960), for group analysis the time depends solely on the number of areas (Crawford and Wishart, 1967, 518). Having made this important point we will leave consideration of individual techniques until the discussion on 'Division and regionalisation' (pp. 37–40). Here we will consider monothetic dichotomous techniques as a group.

Sneath (1962) points out that traditional pre-quantitative classifications are monothetic, as is logical division (Grigg, 1965, 1967). Sokal and Sneath (1963) argue that this was not so much a choice but a necessity before the general availability of computers. However the monothetic approach does have one fundamental logical advantage. In any comprehensive classification covering a large range of taxa it cannot be expected that variables important for classification of the highest order taxa are necessarily those most relevant for lower order taxa. For instance, we would not use the same variables to distinguish mammals from fish as we would, say, to distinguish two species of rodents. Thus we should not use a single discriminant space as in the techniques described in section II above. In regional terms this simply means that the variables relevant at the continental scale are different from those used at an intra-urban level. Whether this argument is valid within the range covered by most research problems in geography seems improbable. Certainly there has been no quantitative attempt to produce a system of regions from continent to urban tract. In fact a single discriminant space is an inherent assumption of most regionalising (e.g. Berry, 1961a). In this situation the choice of either a monothetic or polythetic approach is not immediately obvious given the present state of taxometrics.

b Polythetic division: In general monothetic techniques are equated with division, and polythetic with grouping. We have already described a monothetic grouping procedure (p. 28) (Lockhart and Hartman, 1963); divisive polythetic procedures have been devised, and we now proceed to these.

Although not as critical of monothetic techniques as Sokal and Sneath (1963), McNaughton-Smith (1965, 8) thinks it is 'obviously desirable to have a divisive polythetic method' and the result is *dissimilarity analysis* (McNaughton-Smith *et al.*, 1964). As with most agglomerative methods the first stage is simply to choose some dissimilarity or distance measure. The area which is furthest from the rest of the OTUs is chosen as the first group former and its complement treated as a new population. For each area in the group former's complement the average distance to all other OTUs in the complement is calculated and compared to its distance to the group former. The OTU which has the largest difference between the distances, the average distance being longer, is allocated to the group former. The procedure iterates until average distances within the complement are all shorter than average distances to the group former's set. When the above iteration comes to an end the original population will have been subdivided into two classes, the group and its complement. These classes now constitute new populations and the whole procedure is repeated. Thus it can be seen that the procedure is very much a combination of grouping and division. As well as polythetic McNaughton-Smith *et al.* (1964, 1034) claim that it is 'computationally manageable even on a fairly large scale'. This property cannot be claimed for the technique Edwards and Cavalli-Sforza term 'cluster analysis'.

Whereas the agglomerative method of Lockhart and Hartman (1963) has its main affinities with divisive procedures as was previously noted, the divisive method of Edwards and Cavalli-Sforza (1965) is very closely related to the grouping techniques of section II. The method is quite straightforward, starting with the base areas located in a discriminant space and following the preliminary procedures described in the last section (pp. 31–2). The total population of areas is then divided into two parts in all $2^{n-1} - 1$ possible ways and for each partition an analysis of variance is carried out in the *n* dimensions. The division with the highest 'between sum of the squares' (and consequently smallest 'within sum of the squares') is now chosen as the best split and the two subdivisions of this best split are treated as new populations. The previous steps are repeated until each cluster contains only one point.

The procedure therefore produces a hierarchic tree ranging from complete generality to complete detail like the hierarchic grouping techniques of section II. The main criticism of the method is its comprehensiveness which, as noted in the introduction, requires a great deal of computer time even for small problems.

Another divisive procedure which is closely related to grouping algorithms is McQuitty's (1968) and McQuitty and Clark's (1968) *iterative intercolumnar correlational analysis.* Instead of just considering single links in a correlation matrix this procedure repeatedly correlates columns forming new matrices until a final matrix results consisting of only positive and negative unities. The population is divided on this basis and each submatrix operated on separately in a second cycle, and so on. This approach, which uses an OTU's whole similarity 'profile' as in some early econometric studies (Berry, 1967b, 237), is being experimented with in a geographical context by Johnston (1969).

Of course these hierarchic procedures enjoy the advantages, but suffer from the same disadvantages as their counterparts in grouping strategies. Thus for the same reasons as in the agglomerative approach non-hierarchic methods of division have been developed.

2 Non-hierarchic division

Two types of non-hierarchic division can be identified—univariate and multivariate. The former involves very simple techniques which are briefly described first.

a Univariate division: There are two univariate taxonomic problems that involve division. The first concerns the choosing of class intervals in statistical mapping. Very often this will involve the mapping of factor scores (for example, Kendall, 1939). Although not usually explicitly recognised as a taxonomic problem this can be seen to involve a classification exercise in one dimension. As elsewhere in taxonomy different procedures have been developed for different purposes (Haggett, 1965, 215). The second problem is stratification for sampling on the basis of a single dependent variable. Procedures developed in this context, such as the 'cum $\sqrt{f}$ rule' (Cochran, 1963), are much more closely related to other taxonomic procedures in that they specifically attempt to minimise intragroup variance. However these univariate procedures are in general of limited interest in regional taxonomy and can usually be considered only a preliminary to more complex taxonomic investigation on a multivariate (or multifactor) basis. For this reason they will not be considered further here.

b Multivariate division: There are two non-hierarchic procedures described under this heading. The first introduced by Norman (1968) shows how a universe can be divided into from two to *n* classes with no hierarchic assumptions. In the second case Rose (1964) presents an original contribution which, in purpose at least, is very closely related to the natural grouping algorithms described in section II (pp. 27–8).

The procedure Norman (1968) describes uses simple dichotomous division and departs from the rigid hierarchic tree so that the space is divided, in turn, into from two to *n* classes as specified. The complete algorithm consists of a set of five computer programs. Program I is a preliminary program into which the data is fed and can then be standardised or weighted as required. The space is thus defined in which OTUs can be identified. Program 2 is the actual classification program which we will consider in more detail. The first step is to divide arbitrarily the OTUs into two classes. A variable can have been specified in program I so that OTUs are allocated to two groups—above and below its mean. This monothetic division is now 'improved' in a polythetic manner. The means of each arbitrary group in the discriminant space are computed and any OTU that is closer to the mean of the group it does not belong to is reallocated to that group. Group means are now recalculated and the procedure repeated until every group member is closer to its own group's mean in the discriminant space. Unlike all other post-Goodall (1953) procedures these two resulting classes are not treated as equal ranked hierarchic classes. Instead the class with the larger variance is divided into two new classes about the mean of the variable with the greatest variation in that class. The reallocation process now proceeds with three classes, OTUs being transferred between any of the three groups. It is in this step that the hierarchic structure of the solution will, in all probability, be lost. When all OTUs are again nearest their own group mean the procedure is repeated for four groups, and so on. The program Norman (1968) describes allows up to 50 classes to be produced in this way, although no mention is made of the time involved. Programs 3, and 5 are output programs printing out the results from program 2.

One final point that Norman (1968) makes requires to be mentioned. The procedure of allocation to the nearest mean produces a property of minimised intra-group variance (Howard, 1966), but the results are only locally optimal, as in hierarchic solutions (Scott, 1969a). OTUs are considered for reallocation individually although it might well be possible that the transfer of a particular set of OTUs together might further minimise intra-group variance. In particular the initial subdivision, the first step of program 2 above, will influence the exact nature of the result. Norman's (1968) method is to divide about the mean of the first principal component of a previous analysis of the variables used in the classification. However like most of the procedures described in this paper which attempt to minimise intra-group variance, the solution does not necessarily represent the global optimal.

Norman's procedure has many affinities with most of the previous algorithms described above. This cannot be said of Rose's (1964) very distinctive 'path sampling' approach. First of all Rose defines the problem in terms of graph theory. Graph-theory analysis with interaction data has been described above (Nystuen and Dacey, 1961), but of course this type of approach is just as applicable with similarity links (Harary, 1964). In the graph, OTUs are represented by points and are connected together by lines when they resemble one another above a specified threshold. Given this basis, Rose's method involves finding those links on the graph which are most likely to be in 'natural' divisive positions. Two points are randomly chosen and the shortest path between them along the links is found. The process is repeated many times and each time a path uses a link this is recorded. Using such sample results Rose is able to identify links that are used significantly more times than expected. Such links from 'bridges' between clusters and so are natural cut-off points. Rose mentions experiments he has carried out with this method and the results seem encouraging. Although he emphasises the need for a large sample of paths he does not indicate computer time consumption. None the less the procedure must be considered a valuable addition to taxonomic algorithms not least because it provides a divisive method for detecting 'natural clusters'.

3 Division and regionalisation

In these concluding comments concerning division and regionalising we will first of all consider the hierarchic procedures developed in plant ecology. By way of introduction two unpublished examples of the use of these techniques in geography will be described.

As a preliminary stage in a study of byelaw housing in Hull, Forster (1968) needed to classify the city's cul-de-sac terraces and he accomplished this using 13 morphological elements of the housing in a Williams and Lambert (1959) association analysis. His first division occurred on the cul-de-sac attribute of above or below 20 feet in width. Division continued until the 1% probability level, at which point the original 1,456 cul-de-sacs were arranged in a system of 22 classes.

In a second geographical example Hirst (1968), used ethnic, age and sex data for Tanganyika in a study of in- and out-migration. Using association analysis again, the 554 census areas were first of all divided on the male child/adult ratio variable and each resulting class was divided once more to the 1% level of significance. Thus four migration region types were identified.

These two studies will form the basis of our initial observations concerning division and regionalising, and these will therefore obviously be largely concerned with association analysis.

There are two main points we wish to follow up which arise from Forster's and Hirst's work. The first concerns two aspects of the level of measurement. Forster's study is interesting in that variables at different levels of measurement are involved. An interesting problem arises where absence of one feature *ensures* presence of another. For instance, absence of slate roofing ensures pantile roofing, so that the latter attribute is not included in the analysis. With three exclusive types further problems arise. For instance a house can have one of three types of front window arrangement each mutually exclusive. Presence of one window type ensures the exclusion of two others, so that to include just one type in the analysis would mean information on an important morphological element would be lost. On the other hand inclusion of all three as variables would lead to high χ^2 scores with zero cell values in contingency tables. Forster (1968) compromises by including two of the window types and implying the third by its absence. Thus no information is lost but the validity of the resulting χ^2 values is less certain especially as the error will tend to increase the χ^2 values concerned. This appears to be a typical type of problem that can occur when transferring a technique from one research situation to another and does not seem to have been considered in the ecological literature.

The second measurement problem, however, was foreseen by Goodall (1953) and has been considered in subsequent papers (Williams and Dale, 1965). This concerns the use of association analysis with continuous variables. Both Forster's and Hirst's studies have interval level variables. Forster dichotomises his by using empirical evidence which is supported by the varying byelaw stipulations. All six of Hirst's variables are continuous and they are converted into a binary scale by using divisions about national averages. Thus in both geographical examples some loss of information is accepted so that the association analysis techniques become available for the research problem. This conversion to a binary scale was suggested by Goodall (1953, 43) where he also noted the possibility of using the product moment correlation coefficient. As this is designed for interval data the above examples suggest that the possible replacement of χ^2 by r should be of interest. Williams and Dale

(1965, p. 63) point out that in the 2 × 2 case $\chi^2 = Nr^2$ so that the indicator statistic $\sum \chi^2$ can be regarded as $\sum r^2$. In this form these procedures would seem applicable to an interval level of measurement but as far as we are aware no substantive work on this possible development has yet been published.

While we are considering levels of measurement we can note that the possibility of the complete use of pure presence–absence data in geographical classifications appears to be fairly remote. In the geographical examples cited above some presence–absence attributes are used by Forster but he finds it necessary to include other types of variables in his analysis. As it is also probably true that the geographer will be concerned with fewer variables than the ecologist's species (which tend to be >100), Crawford and Wishart's group analysis would seem at this point to have limited potential in geographical classification.

The second point we wish to make in our comments on the geographical applications of association analysis concerns regionalising *per se* as opposed to the term 'geographical classification' used above. Forster's classification is not concerned with the location of the OTUs so that his solution is a classification of cul-de-sacs and not a regionalisation in any form. Hirst defines 'migration regions' but these are regional types, his areal units being partitioned irrespective of their areal location. In fact it is hard to see how location can be incorporated in these divisive techniques in order to produce contiguous regions. Perhaps if the relevant variables were depicted as isolines a system of regions might be produced, but this must be considered a very tentative suggestion. In this context it was also noted above (III, 2, a) that the statistical map categories only represent univariate regional types. At this point therefore we can turn to the polythetic procedures and consider whether these are more amenable to regionalising.

There are no examples of the use of dissimilarity analysis in geography. None the less it is partly agglomerative in its definition of classes and the possibility of incorporating contiguity constraint at this stage of the procedure may be considered.

In Edwards and Cavalli-Sforza's cluster analysis regionalisation may have distinct advantage over the normal classification it was designed for. The main objection to this method is its computational requirements. However, with regionalising, a contiguity constraint will mean that many of the $2^{n-1} - 1$ possible divisions are not required for comparison, thus greatly easing the computational pressure. The number of valid divisions at each step will depend on the pattern of base areas, but the restrictions of regionalising might mean that this quite comprehensive technique is more useful in this context than in other classifications. As yet, however, there are no examples of the use of this procedure for any classification in geography.

In contrast we should note that the non-hierarchic multivariate divisive procedure of Norman (1968) as well as incorporating advantages over other procedures illustrated above, has been developed in a geographical context. The research problem which this technique was constructed for was the classification of London's census enumeration districts. Using indices derived from the census these districts were classified into just six regional types. It was not found necessary to consider location within the classification although modification of this procedure to produce contiguous regions may be considered a possible development. Thus we can perhaps conclude that this non-hierarchic divisive procedure appears to be potentially the most useful for regionalising *per se*. Rose's (1964) method like other procedures for finding natural clusters seems particularly unsuited to a contiguity constraint.

In general our conclusion for divisive procedures is that these techniques seem to have a more limited use in regional taxonomy than the wider range of agglomerative procedures. In other geographical classifications they can be useful as Forster (1968), Hirst (1968) and Norman (1968) have shown, and they should at least be considered in such problems as town classification where Haggett (1965, 257) reports another application of association analysis (Caroe, 1968).

IV Analysis and testing of developed regional structures

This section is concerned with a group of methods which analyse or test previously developed regional systems. A regional structure is given and the problem is to test whether it is satisfactory with respect to certain specifications. The assignment problem in regionalisation is a familiar example of this nature. Here three or more regions characterised by specific phenomena are given, one of which has to be assigned to the most similar remaining region. Assignment is made objective by the use of a significance test, two examples of which—chi-square tests and variance analysis—are presented here. The use of discriminant analysis seems to be a relatively little-known area of research in this field and the way in which it analyses and tests predeveloped regional structures will also be outlined.

1 Statistical tests of significance

The first approach involves the transfer of techniques from the task for which they were designed, namely the testing of observed differences between samples, to the evaluation of specified regional arrangements of areal units. The use of statistical tests for such problems was pioneered by Zobler (1957, 1958a). A prime assumption of this approach is that the data be considered as a sample even if they appear to constitute a finite population. Zobler suggests this theoretical defect is overcome if one thinks of the population under study as a sample of a hypothetical larger universe, such as the infinite number of areal divisions of a hypothetical larger universe or the infinite number of areal divisions of a nation. If one accepts this argument, and Gould (1969) does not, then the whole range of statistical significance tests are potentially available for investigating regional structures. Two such tests have been suggested and illustrated by Zobler and others.

a Chi-square test approach: An interesting debate was initiated in the late nineteen-fifties on the uses and misuses of chi-square in region construction (Zobler, 1957, 1958a, 1958b; Berry, 1958, 1959; Mackay, 1958, 1959). The problem is to decide whether areas are significantly similar or different from one another on the basis of certain attributes. These attributes must be quantified in an absolute frequency form, i.e. non-relative in that they cannot be expressed in other units of measure. Chi-square is a direct way of comparing two distributions—the observed and the expected or theoretical frequencies of attributes. The expected frequencies are calculated under the null hypothesis that the two areas under study are perfectly homogeneous, and hence are weighted in proportion to the total attribute population in each area. The null hypothesis is tested by chi-square in the usual manner. Then according to the number of degrees of freedom calculated from the range of independent observations and the level of probability, the hypothesis and also the boundary lines between areas can be accepted or rejected.

In one study Zobler (1958a) examines the problem of the best allocation of West Virginia to one of 3 groups of states (Mid-Atlantic States, South Atlantic States and East-South Central States) on the basis of the number of workers in the primary and secondary sectors of the economy. Chi-square was used to test the observed distribution of the labour force for each area (with the expected distribution) calculated under the null hypothesis that there is no significant difference between the regions. Tests were made for the 3 possible areal combinations (W.Va./M.A., W.Va./S.A. and W.Va./E.S.C.) and it was found that only in the last case were there non-significant differences in the distributions. As a result

West Virginia was combined with the East-South Central States. In another study (controversial in that it employed relative frequencies) Zobler (1957) used a similar method to test a set of physical regions with land use, soil, type of farm and population data in Salem County, New Jersey.

An important operational problem in the use of chi-square in regionalisation concerns the necessity to test all possible pairs of areal units (for true regionalisation these must be contiguous). Alternatively, a partly subjective approach could be used. Groups of 'core' areas could be demarcated empirically and the peripheral 'doubtful' areas could be tested.

A more important defect of chi-square is that it is not an efficient discriminant technique when high frequencies are used, and when there are several possible alternative regionalisations. Not unnaturally this is a well recognised limitation in the statistical literature. It was indicated in the original paper on the test (Pearson, 1900) and by subsequent statisticians (Cochran, 1952, 1954). For reasonable probability levels, as many regions as there are original areal units would often result. This was in fact the case in several class exercises set by one of the present authors using employment data for groups of British counties. Indeed the choice of probability level is a subjective judgement, and is of paramount importance in the final regionalisation. As a corollary, this method does not give any indication of 'best' or optimum regionalisations. Thus, as Zobler has pointed out, the use of chi-square is best 'for assigning unknown unit areas to regions with which they are homogeneous'. This is the assignment problem in the analysis of developed regional structures.

b Variance analysis approach: Coincident with the use of chi-square was the application of variance analysis to regionalisation (Zobler, 1958a). Variance analysis is a technique used here to maximise differences between, and to minimise differences within, regions on the basis of certain attributes. This is operationalised using Snedecor's variance ratio test represented by the statistic F. Snedecor's F is the ratio of the sums of squares of the two types of variation (intra-regional and inter-regional), each having been divided by their respective degrees of freedom. A null hypothesis is set up that there is no significant difference between the regions represented by their attributes, and the test applied in the usual manner. Then, according to the level of probability chosen, and the number of degrees of freedom of the within-region and between-region variances, the hypothesis can be accepted or rejected. If the critical F-value is exceeded the hypothesis is rejected indicating that the tested regionalisation has significantly different internal regional variation from inter-regional variation. If it is not exceeded then the null hypothesis is accepted and the regional construct is not validated.

Variance analysis regionalisation was applied by Zobler (1958a) to the same data previously described as an example of the use of chi-square. F-tests were made for each combination of states, West Virginia being allocated to each group in turn. Significant results were obtained for regionalisations of West Virginia with the East-South-Central States and the South Atlantic States but that with the Mid-Atlantic States was not significant. The first two regionalisations can be compared as they have the same levels of probability and degrees of freedom. It was found that the greatest relative degree of association between the regionalisation and the areal distribution of the labour force was when West Virginia was grouped with the East-South-Central States. Hence the analysis of variance supports for this example the chi-square findings in a more rigorous way.

Another use of the analysis of variance, which can be contrasted with that above, occurs in social area analysis. Here the problem is to test certain hypotheses about the spatial variation of certain variables in the city. The hypotheses are related to the established or classical theoretical notions of concentric zones after Burgess and sectors after Hoyt. In a study by Anderson and Egeland (1961), 16 census tracts were selected at random from four American cities. Each census tract was characterised by certain indices—for example, a prestige value index—and by an urbanisation index, and selected from one of four 30° sectors and from one of four zones at progressive distances from the city centre. The first hypothesis to be tested was that there was no significant difference between sectors. Because of variation in orientation of sectors (cities could have their characteristic sectors in varying places) each city was considered separately. The hypothesis was rejected at the 0·01 level for all four cities, Akron, Dayton, Indianapolis and Syracuse. This meant that the sectors were significantly different and so in a social area analysis they could be separated. The second hypothesis to be tested on this particular variable was that there was no significant difference between zones. Again each city was considered separately and the hypothesis was accepted for three of the cities, Akron, Dayton and Syracuse. This meant that there was no difference between zones and so in a social area analysis there would be no reason to consider zones in isolation. The hypothesis was rejected for Indianapolis although the sectoral variation was much more important. The cities were all considered together on the urbanisation index. The null hypotheses relating to variation among sectors and zones were respectively accepted and rejected. Thus for this index variation among zones was the more important, and the social areas in terms of zones were verified.

In a similar study by Rees (1968) social areas were again tested by this method. Here the indices used were scores on the first two factors of a principal components analysis of socio-economic variables, namely socio-economic status and family status. In this study of Chicago, it was found that there were significant differences between zones and sectors on both indices. In other words, the theoretical notions of both Burgess and Hoyt were both borne out in this study, which is not surprising as it was in Chicago that these two urban investigators did their major work. Similar studies have been undertaken by Murdie (1968) in Metropolitan Toronto, and by McElrath and Barkey (1964), cited in Rees (1968).

A comparison of F-values for a series of different regionalisations enables the choice of the maximum inter-regional differential to be made. So, as with chi-square, the use of variance analysis in a large study would require the testing of numerous pairs of areal units. Again, similar criticisms as to the use of chi-square, concerning the subjective judgement involved in the choice of probability levels, and the absence of measures of efficiency in regionalisation can be made. In addition as was pointed out above, in the transfer of these statistical tests to the testing of regional structures it usually has to be assumed that the data represent a sample from some hypothetical larger universe. The use of analysis of variance, however, involves further assumptions as the technique is parametric in nature. Thus one has to assume that the hypothetical universe not only exists but does so in a specified form—the normal frequency distribution. To make this further assumption may seem unrealistic, so that a non-parametric equivalent to the Snedecor test may be considered preferable. The Kruskal Wallis one-way analysis of variance is one such technique (Siegel, 1956). Although information is lost through the use of an ordinal scale of measurement the technique has a power efficiency of 95·5% compared with the F-test. This loss has to be set against the avoidance of parametric conditions. This technique has been used in this context by Johnston (1965).

As an illustration of the technique in this regional context the data that Zobler (1958a) used to illustrate the F-test have been reworked. The Kruskal Wallis H statistic was calculated for each assignment of West Virginia, i.e., to the Mid-Atlantic region, to the South Atlantic region, and to the East-South-Central region, in turn. Respectively the values for H were 11·89, 12·64 and 12·98. With two degrees of freedom the probability of the null hypothesis of no differences between regions can be stated: $0{\cdot}01 < H < 0{\cdot}001$ for each regional arrangements of states. Thus on this basis any of the three regional structures hypothesised can be reasonably considered statistically significant.

These results, while largely agreeing with those of the Zobler study, highlight the fundamental limitation of using significance tests for investigating regional structures—the fact that the techniques are designed to investigate and test single situations. Thus when used in the manner described above they may give several regional arrangements which are statistically highly significant while not necessarily optimal, as Garrison has pointed out in a footnote to Zobler's paper. Of course as Zobler mentions one can use the *F* ratio (or *H* statistic) to indicate which of the regional arrangements tested is the most significant, but to find the best solution for these methods would involve testing all possible regional combinations.

However such a cumbersome procedure involving innumerable statistical tests on 'pseudo-samples' is superfluous with the availability of the multiple discriminant analysis approach. Thus the statistical significance test approach initiated by Zobler has not been followed up in the regionalisation literature. It can be considered of value for the simple assignment problem, but for further analysis of regional structures it has been largely superseded by more powerful techniques such as discriminant analysis which have been developed specifically for the classification problem.

2 Multiple discriminant analysis

A second example of the analysis of developed regional structures concerns the application of discriminant analysis. In 1964 two important papers dealing with multiple discriminant functions appeared in a series concerned with the study of computer applications in the earth and environmental sciences (Casetti, 1964a, 1964b). More specifically the papers deal with the application of discriminant analysis to classificatory and regional analysis.

Given a classification (regionalisation) of items (regions) measured over a range of attributes (variables), a linear discriminant function consists of a linear combination of the variables which yield an optimal discrimination of the items in a classification. Such functions can be used in discriminant procedures that define rules for the assignment of a new item to an existing class. Usually these rules assign new items on the basis of 'nearness' in discriminant space. Linear discriminant functions themselves give the dimensions of classifications in a similar way as principal components analysis gives dimensions of variability. Also the original variables' discriminating power is distributed among the functions in much the same way as variability is distributed over the principal components.

Casetti gives two examples of the use of multiple discriminant functions. The first, an analysis of the classification of Italian regions, used as input the first six orthogonal basic patterns of a principal components analysis of 25 socio economic demographic variables. The study was based on the traditional division of Italian regions into North, Central and South.

The North is made up of Lombardia, Piemonte, Valle d'Aosta, Trentino Alto Adige, Veneto, Fruili-Venezia Giulia, Liguria and Emilia-Romagna. Toscana, Umbria, Marche, Lazio, Abruzzi and Molise comprise the Central Region. Lastly the South is made up of Campania, Puglia, Basilicata, Calabria, Sicilia and Sardegna. By means of the linear discriminant function program, documented in the paper, 95% of the inter-class variability was accounted for by the first linear discriminant function. This figure rose to nearly 100% with the first two functions, which are essentially a measure of the discriminatory power of the classification. As with principal components analysis the loadings and scores of the variables and observations, in this case principal components and Italian regions, can be evaluated with respect to the linear discriminant function. Casetti concludes that the three-fold classification is accurate and is based on one main dimension—that of economic and social development.

The second example of discriminant analysis of classifications concerns the climatic regions of North America. The first six orthogonal basic patterns of a principal components analysis of 24 temperature–precipitation variables were used as data input. 70 weather stations in North America were assigned in turn to one of eight climatic types (a modified Köppen classification—B. Caf. Cbf. Cs. Daf. Dbf. Def. Et.). From the discriminant analysis two basic dimensions for the classification were evolved in two linear discriminant functions accounting for 66% and 34% of the variability respectively. The first dimension illustrates that climatic character at the extremes of which are the winter rains of the west coast and the summer temperature characteristics of the east coast. The second is another complex dimension ranging from subzero temperatures and low precipitation of the north and interior of the continent to the summer rains of the coasts. By plotting the scores of each station with respect to the first two linear discriminant functions the classification can be evaluated. The first function isolates each of the three *C* climates very well in a trend from north-west to south-east. The second function best describes three clusters Def. Et; B. Dbf; Daf. Caf. Cs. Cbf.—in the north interior, the central or 'core' parts of the continent, and the coastal belts respectively.

Casetti's second paper deals more specifically with regionalisation as opposed to the regional analysis above. It builds on methods previously described, and introduces the technique of discriminant iterations. Discriminant iterations use discriminant procedures, where the latter consist of certain basic rules for assigning new objects to an existing classification. These rules can be varied according to the type of classification required: for example, various distance functions can be used to measure similarity of points in the discriminant space. Each discriminant iteration allocates the new object to its respective class according to the results of its discriminant function. A new classification is produced on which another iteration can be performed, and so on until the optimum classification is produced.

The same North American climatic data set as used previously (see p. 46) illustrates the discriminant iterations method. The input to the iterations computer program, documented in the paper, were the first six orthogonal basic patterns of a principal components analysis of the 24 variables over 70 meteorological stations. Three classifications were analysed each having more and more random noise—an eight group modified Köppen classification, a 'disordered' classification (removal and random replacement of the subset cards), and a 'shuffled' classification (removal, shuffle of data cards, and replacement of subset cards). The iterations program tests the quality of each classification with respect to the data set and to the desired discriminant procedures by developing a limit or optimal classification. The cores of classes are then identified and the two classifications, before and after iteration, are compared to determine, by means of chi-square tests, whether the differences are due to random elements or not. Results show that the Köppen classification was nearer to its limit in terms of iterations than the random classifications. Also the disordered classification appeared to be 'better' than the shuffled classification in this respect. With each iteration a measure of improvement in terms of group homogeneity or clustering is given, and in this way each classification is further improved to near optimality. Indeed on the final iteration all three classifications show remarkably similar traits, especially when mapped.

Another example of the use of this method to improve existing classifications is given by Spence (1969a), who analysed a hierarchical system of employment classification of English counties made in 1961. The predeveloped typology was produced by the usual methods of principal components analysis and distance grouping procedures applied to the county observation by employment variable data matrix (Spence, 1968). The paper used Demirmen's program (1969), which is a revised and extended version of the discriminant iterations procedure of Casetti, to examine ways in which classifications that have been extracted from the hierarchical system are sub-optimal. The hierarchical grouping method of Ward (1963) has been criticised (Scott, 1969a) on the grounds that it cannot overcome the possibility of developing and building upon a local sub-optimum solution on which it might have converged. The question arose as to which level of the hierarchical grouping or linkage tree to analyse. As one could not reasonably expect the whole system of classifications to be uniformly sub-optimal, several levels were examined. In essence the study indicated that for the particular empirical example under study the hierarchy of classifications was very near optimality.

Only one item was incorrectly grouped by the Ward procedure and this occurred in the stage of the hierarchy where there were many groups. This study was one of classification *per se*, not regionalisation. Work by one of us is in hand to add a contiguity constraint to Casetti's program in order that an analysis of a true hierarchy of regional systems can be undertaken.

In addition to their other geographical applications (Ahmad, 1965), discriminant functions have been extensively used as a classification procedure in several related fields, notably in geological taxonomy. Geological applications include differentiating oil- and ore-bearing rocks from barren ones, classification of refractory and non-refractory quartzites (among several stratigraphic and palaeontological classifications), and the geographically interesting regionalisation of areas of high mineral exploration potential. These and several more examples are described further in Griffiths (1966), Miller and Kahn (1962) and Krumbein and Graybill (1965) together with a mathematical discussion of the discriminant function statistical model in the case of the last two works. In addition to Casetti's computer programs the work of Davis and Sampson (1966) should be noted, especially with respect to the programming of the smaller computers (Davis, 1966).

This discussion of discriminant analysis, which can only analyse, test and improve existing classifications/regionalisations, concludes the section on analysis and testing of developed regional structures.

V Concluding remarks

This paper can be viewed as an expanded version of two previously unpublished discussion papers (Spence, 1967; Taylor, 1968). However despite the large expansion it remains virtually impossible to achieve complete comprehensiveness in such a diverse and fast-growing field as taxometrics. This point is emphasised when the references cited in this paper are compared to those of other reviews, such as Sokal and Sneath (1963), Lance and Williams (1967, 1968a), Jones (1968) and especially Ball (1965), where there is surprisingly little cross-referencing considering all the reviews purport to consider the same basic problem. Possibly the most complete bibliography is the 'KWIC index to Taxometrics (Numerical Taxonomy) literature' (Maisel, 1969). This has been prepared at the Computation Center, Georgetown University, and comprises the bibliographic sections of Nos. 1 to 12 inclusive of the newsletter *Taxometrics*. It is perhaps of interest that although this newsletter attempts to cover the whole range of quantitative taxonomy it has so far wholly neglected the literature in geography. Thus this paper has come full circle from the initial comments concerning the neglect of the taxonomic literature by geographers. It would therefore seem appropriate to conclude with some comments concerning the analogy between classification and regionalisation which we have previously identified as the basic assumption of this paper.

The discussion so far has generally followed the traditional distinction between contiguous regions and non-contiguous regional types (Grigg, 1965; Berry, 1965) or homogeneous spaces (Boudeville, 1966; Berry, 1968). However two recent papers cast doubt upon this distinction. Johnston (1969), following Czyz (1968), argues that the definition of homogeneous regions is a two-stage procedure involving first an algorithm to arrange areal units into classes or regional types, and then inspection of their distribution to identify regions *per se*. Another example of this procedure is the work on the regionalisation of Pennsylvania counties for development planning by Stevens and Brackett (1968). Taylor and Spence (1969) make two criticisms of this procedure. First of all the procedure will usually mean that a system with only a small number of regions cannot be produced. Thus a division of several hundred OTUs into just two classes, the most general classification possible, will still be likely to produce a large number of regions. The work of Gittus (1964–5) who analyses several hundred census enumeration districts in Hampshire and Merseyside can be cited as an example of just such a complex regional solution. The second criticism follows from the first in that such a regionalisation into, say, 15 regions from just two classes will produce very heterogeneous regions for a scale of 15. Reference to the results presented by Spence (1968, Table 12, Figures 7 and 8) comparing, for instance, the 16 regions produced without a contiguity constraint with the set of 17 built up under contiguity amply confirms this second point. These arguments are expanded in Taylor and Spence (1969).

The counter argument used by Johnston (1969) involves citing the need for more 'natural' areal classes. It has already been noted that in both grouping and division the usual algorithms will not necessarily identify natural clusters in the sense of Simpson (1951), although some procedures are available for this purpose. It has also been previously pointed out that the imposition of a constraint on these particular algorithms is inappropriate in respect to their aims. Quite obviously the argument now reverts to whether contiguous regions, 'natural' clusters or some other systems of classes are relevant to the particular research context in which the taxonomic exercise is taking place. This inevitably brings us back to the point emphasised in the introduction, namely that the purpose to which the classification is to be put is of paramount importance. Thus if contiguous groups are required for the purpose of the overall research project then it is not only legitimate but, bearing in mind the inefficiency of the Johnston–Czyz two-stage proposal, it is also essential to incorporate a contiguity constraint. Similarly 'compact' or 'equal area' constraints may be imposed just as in biological taxonomy a 'discrete constraint' has always been implicitly incorporated. Thus the conclusion is that there is no reason why any constraint should not be incorporated in a taxonomic algorithm if the research context requires it, and in geography this will often involve a contiguity constraint.

VI References

Abiodun, J. O. 1968 : Central place study in Abeokuta province, South-western Nigeria. *Journal of Regional Science* 8, 57–76.

Abu-Lughod, J. N.D. : The factorial ecology of Cairo. Northwestern University, Department of Geography : unpublished paper.

Ahmad, Q. S. 1965 : Indian cities : characteristics and correlates. *University of Chicago, Department of Geography, Research Paper* 102.

Anderson, T. R. and **Egeland, J. A.** 1961 : Spatial aspects of social area analysis. *American Sociological Review* 26, 392–8.

Austin, M. P. and **Orlocci, L.** 1966 : Geometric models in ecology. II : an evaluation of some ordination techniques. *Journal of Ecology* 54, 217–27.

Ball, G. H. 1965 : Data analysis in the social sciences : what about the details? In *1965 Joint Conference Proceedings, American Federation of Information Processing Society.*

Bassett, K. and **Downs, R. M.** 1968 : *The use of multidimensional spatial models in geographical research.* University of Bristol, Department of Geography : unpublished manuscript.

Bell, W. 1953 : The social areas of the San Francisco Bay region. *American Sociological Review* 18, 39–47.

Berry, B. J. L. N.D. : American urban dimensions, 1960. University of Chicago, Centre for Urban Studies : unpublished paper.

1958 : A note concerning methods of classification. *Annals of the Association of American Geographers* 48, 300–3.

1959 : Comments on the use of chi square. *Annals of the Association of American Geographers* 49, 89.

1960 : An inductive approach to the regionalization of economic development. In Ginsberg, N., editor, Essays on geography and economic development, *University of Chicago, Department of Geography, Research Paper* 62, 78–107.

1961a: A method for deriving multifactor uniform regions. *Przegląd Geograficzny* 33, 263–82.
1961b: Basic patterns of economic development. In Ginsberg, N., editor, Atlas of economic development, *University of Chicago, Department of Geography, Research Paper* 68, 110–19.
1964: Approaches to regional analysis: a synthesis. *Annals of the Association of American Geographers* 54, 2–11.
1965: Identification of declining regions: an empirical study of the dimensions of rural poverty. In Wood, W. D. and Thomas, R. S., editors, *Areas of Economic Stress in Canada*, Kingston, Ontario, 22–66.
1966: Essays on commodity flows and the spatial structure of the Indian economy. *University of Chicago, Department of Geography, Research Paper* 111.
1967a: The mathematics of economic regionalization. *Proceedings of the Fourth General Meeting of the IGU Commission on Methods of Economic Regionalization, 1965, Brno, Czechoslovakia*, Prague, 77–106.
1967b: Grouping and regionalizing: an approach to the problem using multivariate analysis. In Garrison, W. L. and Marble, D. F., editors, *Quantitative geography, Part I: economic and cultural topics, Northwestern University, Studies in Geography* 13, 219–51.
1968: Numerical regionalization of political economic space. In Berry, B. J. L. and Wróbel, A., editors, 1968, 27–35.

Berry, B. J. L. and **Hankins, T. D.** 1963: A bibliographic guide to the economic regions of the United States. *University of Chicago, Department of Geography, Research Paper* 87.

Berry, B. J. L. and **Rees, P. H.** 1969: The factorial ecology of Calcutta. *American Journal of Sociology* 74, 445–91.

Berry, B. J. L and **Wróbel, A.** editors, 1969: *Economic regionalization and numerical methods.* Final report of the Commission on Methods of Economic Regionalization of the International Geographical Union, *Geographia Polonica* 15.

Blumenthal, L. M. 1953: *Theory and application of distance geometry.* Oxford: Clarendon Press.

Bonner, R. E. 1964: On some clustering techniques. *IBM Journal of Research and Development* 8, 22–32.

Broadbent, T. A. 1968: An introduction to factor analysis and its application in regional science. London: *Centre for Environmental Studies, Working Paper* 13.

Boudeville, J. R. 1966: *Problems of regional economic planning.* Edinburgh University Press.

Bunge, W. 1966a: Theoretical geography. *Lund Studies in Geography, Series C, General and Mathematical Geography* 1, Lund: Gleerup. (Second revised and enlarged edition.)

1966b: Locations are not unique. *Annals of the Association of American Geographers* 56, 375–6.

1966c: Gerrymandering, geography and grouping. *Geographical Review* 56, 256–63.

Carey, G. W. 1966: The regional interpretation of Manhattan population and housing patterns through factor analysis. *Geographical Review* 56, 551–70.

Caroe, L. 1968: A multivariate grouping scheme: association analysis of East Anglian towns. In Bowen, E. G., Carter H. and Taylor, J. A., editors, *Geography at Aberystwyth*, Cardiff, University of Wales Press, 253–69.

Casetti, E. 1964a: Multiple discriminant functions. *Northwestern University Department of Geography, Technical Report* 11, ONR task 389–135, Contract nonr 1228 (26).

1964b: Classificatory and regional analysis by discriminant iterations. *Northwestern University, Department of Geography, Technical Report* 12, ONR task 389–135, Contract nonr 1228 (26).

Caswell, B. 1968: The urban ecology of Miami, Florida. University of Chicago, Centre for Urban Studies: unpublished paper.

Cattell, R. B. 1944: A note on correlation clusters and cluster search methods. *Psychometrika* 9, 169–84.

1965: Factor analysis: an introduction to essentials, Part I: the purpose and underlying models; Part II: the role of factor analysis in research. *Biometrics* 21, 190–215, 405–35.

1966a, editor: *Handbook of multivariate experimental psychology*. Chicago: Rand McNally.

1966b: The data box: its ordering of total resources in terms of possible relational systems. In Cattell, R. B., editor, 1966a.

Cattell, R. B., Coulter, M. A. and **Tsujioka, B.** 1966: The taxonometric recognition of types and functional emergents. In Cattell, R. B., editor, 1966a.

Chisholm, M. 1964: Problems in the classification and use of farming-type regions. *Transactions of the Institute of British Geographers* 35, 91–103.

Claval, P. and **Juillard, E.** 1967: *Région et régionalisation dans la geographie française et dans d'autres sciences sociales*. Paris: Librairie Dalloz.

Clements, F. E. 1954: The use of cluster analysis with anthropological data. *American anthropologist* 56, 180–99.

Cliffe-Phillips, G., Mercer, J. and **Yeung, Y. M.** 1968: The spatial structure of urban areas: a case study of the Montreal metropolitan area. University of Chicago, Centre for Urban Studies: unpublished paper.

Cochran, W. G. 1952: The χ^2 test of goodness of fit. *Annals of Mathematical statistics* 23, 315–45.

1954 : Some methods for strengthening the common χ^2 tests. *Biometrics* 10, 417–51.

1963 : *Sampling techniques*. New York : Wiley. (Second edition.)

Cohen, Y. 1968 : The urban ecology of Birmingham, Alabama. University of Chicago, Centre for Urban Studies : unpublished paper.

Cole, L. C. 1949 : The measurement of interspecific association. *Ecology* 30, 411–24.

1957 : The measurement of partial interspecific association. *Ecology* 38, 226–33.

Cooley, W. W. and **Lohnes, P. R.** 1962 : *Multivariate procedures for the behavioral sciences*. New York : Wiley.

Cooper, L. 1967 : Solutions of generalized locational equilibrium models. *Journal of Regional Science* 7, 1–18.

Cox, D. R. 1957 : Note on grouping. *Journal of the American Statistical Association* 52, 543–7.

Crawford, R. M. M. and **Wishart, D.** 1967 : A rapid multivariate method for the detection and classification of groups of ecologically related species. *Journal of Ecology* 55, 505–24.

Czyż, T. 1968 : The application of multi factor analysis in economic regionalization. In Berry, B. J. L. and Wróbel, A. editors, 1968, 115–34.

Dacey, M. F. 1958 : Nearest neighbor linkage analysis. University of Washington, Department of Geography : unpublished paper.

Davis, J. C. 1966 : Programming the discriminant classification function for small computers. In Merriam D. F. editor, Computer applications in the earth sciences : colloquium on classification procedures, *University of Kansas, State Geological Survey, Computer Contribution* 4, 1–4.

Davis, J. C. and **Sampson, R. J.** 1966 : Fortran II program for multivariate discriminant analysis using an IBM 1620 computer. *State Geological Survey, University of Kansas, Computer Contribution* 4.

Dear, M. J. 1969 : Multivariate analysis of urban structure. *University College London, Department of Town Planning, Discussion Paper* 4.

Demirmen, F. 1969 : Multivariate procedures and Fortran IV program for evaluation and improvement of classifications. *University of Kansas, State Geological Survey, Computer Contribution* 31.

Dziewoński, K. 1968 : Economic regionalization : a report of progress. In Berry, B. J. L. and Wróbel, A., editors, 1968, 9–23.

Eades, D. C. 1965 : The inappropriateness of *r* as a measure of taxonomic resemblance. *Systematic Zoology* 14, 98–100.

Edwards, A. W. F. and **Cavalli-Sforza, L. L.** 1963 : A method for cluster analysis. *Biometrics* 21, 362–75.

Festinger, L., Schacter, S. and **Back, K.** 1955 : Matrix analysis of group structures. In Lazarsfeld, P. F. and Rosenburg, M., editors, *The language of social research*, New York : Free Press, 358–67.

Fisher, R. A. 1936: The use of multiple measurements in taxonomic problems. *Annals of Eugenics* 7, 179–88.

Fisher, W. D. 1958: On grouping for maximum homogeneity. *Journal of the American Statistical Association* 53, 789–98.

Forster, C. A. 1968: The development of byelaw housing in Kingston-upon-Hull: an example of multi-variate morphological analysis. *IBG Study Group in Urban Geography, Salford conference, September, 1968.*

Forsyth, E. and **Katz, L.** 1946: A matrix approach to the analysis of sociometric data: preliminary report. *Sociometry* 9, 340–7.

Friedman, H. P. and **Rubin, J.** 1967: On some invariant criteria for grouping data. *Journal of the American Statistical Association* 62, 1159–1178.

Garrison, W. L. and **Marble, D. F.** 1963: Factor analytic study of the connectivity of a transportation network. *Regional Science Association, Papers and Proceedings* 12, 231–8.

Gilbert, E. W. 1960: The idea of the region. *Geography* 45, 157–75.

Gittus, E. 1964–5: An experiment in the identification of urban sub-areas. *Transactions of the Bartlett Society, University College, London* 2, 109–35.

Goddard, J. B. 1968: Multivariate analysis of activity patterns in the city centre: a London example. *Regional Studies* 2, 69–85.

1970: Functional regions within the city centre: a factor analytic study of taxi flows in Central London. *Transactions of the Institute of British Geographers* 49, 161–82.

Goodall, D. W. 1953: Objective methods for the classification of vegetation: I. The use of positive interspecific correlation. *Australian Journal of Botany* 1, 39–63.

1954: Objective methods for the classification of vegetation: III. An essay in the use of factor analysis. *Australian Journal of Botany* 2, 304–24.

1964: A probabilistic similarity index. *Nature* 203, 1098.

1966: Deviant index: a new tool for numerical taxonomy. *Nature* 210, 216.

Gould, P. R. 1967a: Structuring information on spacio-temporal preferences. *Journal of Regional Science* 7, 259–74.

1967b: On the geographical interpretation of eigenvalues. *Transactions of the Institute of British Geographers* 42, 53–86.

1968: Review of Berry, B. J. L. 1966. *Geographical Review* 63, 158–61.

1969: Is *statistix inferens* the geographical name for a wild goose? *IGU Commission on Quantitative Methods, London, August, 1969.*

Gregory, C. L. 1949: Advanced techniques in the delineation of rural regions. *Rural Sociology* 14, 59–63.

Greig-Smith, P. 1964: *Quantitative plant ecology*. London: Butterworths. (Second edition.)

Griffiths, J. C. 1966: Application of discriminant functions as a classification tool in the geosciences. In Merriam, D. F., editor, Computer applications in the earth sciences: colloquium on classification procedures, *University of Kansas, State Geological Survey, Computer Contribution* 7, 48–52.

Grigg, D. B. 1965: The logic of regional systems. *Annals of the Association of American Geographers* 55, 465–91.

1967: Regions, models and classes. In Chorley, R. J. and Haggett, P., *Models in geography*, London: Methuen, 461–507.

Haggett, P. 1965: *Locational analysis in human geography*. London: Arnold; New York: St Martin's Press.

Hagood, M. J. 1943: Statistical methods for the delineation of regions applied to data on agriculture and population. *Social Forces* 21, 288–97.

Hagood, M. J., Danilevsky, N. and **Beum, C.O.** 1941: An examination of the use of factor analysis in the problem of sub regional delineation. *Rural Sociology* 6, 216–33.

Hagood, M. J. and **Price, D. O.** 1952: *Statistics for sociologists*. New York: Henry Holt.

Harary, F. 1964: A graph theoretical approach to similarity relations. *Psychometrika* 29, 143–51.

Harman, H. 1961: *Modern factor analysis*. Chicago University Press.

Hartigan, J. A. 1967: Representation of similarity matrix by trees. *Journal of the American Statistical Association* 62, 1140–58.

Hartshorne, R. 1939: *The nature of geography*. Chicago and Lancaster.

Hattori, K. K., Kagaya, K. and **Ianaga, S.** 1960: The regional structure of surrounding areas of Tokyo. *Geographical Review of Japan* 33, 495–514.

Henshall, J. D. 1966: The demographic factor in the structure of agriculture in Barbados. *Transactions of the Institute of British Geographers* 38, 183–95.

Henshall, J. D. and **King, L. J.** 1966: Some structural characteristics of peasant agriculture in Barbados. *Economic Geography* 42, 74–84.

Hess, S. W. and **Weaver, J. B.** 1965: Non partisan political redistricting by computer. *Operations Research* 13, 998–1006.

Hirst, M. 1968: Migration regions in Tanganyika, 1957. *IBG Study Group in Population Geography, Salford, September, 1968.*

Holloway, J. D. and **Jardine, N.** 1968: Two approaches to zoogeography: a study based on the distributions of butterflies, birds and bats in the Indo-Australian area. *Proceedings of the Linnean Society, London* 179, 153–88.

Holzinger, K. J. 1937: *Student manual of factor analysis*. University of Chicago Press.

Howard, N. 1966: Classifying a population into homogeneous groups.

In Lawrence, J. R., editor, *Operational research and the social sciences*, London: Tavistock Publications, 585–94.

Huxley, J. S., editor, 1940: *The new systematics*. Oxford: Clarendon Press.

Hyvarinen, L. 1962: Classification of qualitative data. *BIT*, (*Nordisk Tidskrift for Informations-Behandling*) 2, 83.

IGU Commission on Methods of Economic Regionalization, 1962: *Economic regionalization*. Materials of the first general meeting of the commission. *Dokumentacja Geograficzna* 1.

1964: *Methods of Economic Regionalization*. Materials for second meeting of the commission. *Geographia Polonica* 4.

1965: *Aims of Economic Regionalization*. Proceedings of the third general meeting of the commission. *Geographia Polonica* 8.

1968: *Regionalization and Development*. Proceedings of a meeting held in Strasbourg, July, 1967. Paris: Colloques Internationaux du CNRS Sciences Humaines.

Inger, R. F. 1958: Comments on the definition of genera. *Evolution* 12, 370–84.

Isard, W. 1960: *Methods of regional analysis*. New York: Wiley.

James, P. E. 1943: A regional division of Brazil. *Geographical Review* 32, 493–5.

Jancey, R. C. 1966: Multidimensional group analysis. *Australian Journal of Botany* 14, 127.

Jardine, C. J., Jardine, N. and **Sibson, R.** 1967: The structure and construction of taxonomic hierarchies. *Mathematical Biosciences* 1, 173–9.

Jardine, N. and **Sibson, R.** 1968: The construction of hierarchic and non-hierarchic classifications. *Computer Journal* 17, 177–84.

Jeffrey, D., Casetti, E. and **King, L.** N.D.: Economic fluctuations in a multi regional setting: a bi-factor analytic approach. Ohio State University, Department of Geography: unpublished paper.

Johnson, S.C. 1967: Hierarchical clustering schemes. *Psychometrika* 32, 241–54.

Johnston, R. J. 1965: Multi-variate regions: a further approach. *Professional Geographer* 17, (5), 9–12.

1968: Choice in classification: the subjectivity of objective methods. *Annals of the Association of American Geographers*, 58, 575–89.

1969: Grouping and regionalizing: some methodological and technical observations. *IGU Commission on Quantitative Methods, Ann Arbor and London, August, 1969*.

Jones, K. J. 1968: Problems of grouping individuals and the method of modality. *Behavioral Science* 13, 496–511.

Jones, K. S. and **Jackson, D.** 1967: Current approaches to classification and clump finding at the Cambridge Language Research Unit. *Computer Journal* 10, 29–37.

Kendall, M. G. 1939: The geographical distribution of crop productivity in England. *Journal of the Royal Statistical Society* 102, 21–62.

Kershaw, K. A. 1964: *Quantitative and dynamic ecology*. London: Arnold.

Kimble, G. H. T. 1951: The inadequacy of the regional concept. In Stamp, L. D. and Wooldridge, S. W., editors, *London essays in geography*, London, 151–74.

King, B. F. 1966: Market and industry factors in stock price behaviour. *Journal of Business* 39, 139–90.

1967: Step-wise clustering procedures. *Journal of American Statistical Association* 62, 86–101.

King, L. J. 1966: Cross sectional analysis of Canadian urban dimensions: 1951 and 1961. *Canadian Geographer* 10, 205–24.

1969: *Statistical analysis in geography*. Englewood Cliffs: Prentice Hall.

Krumbein, W. C. and **Graybill, F. A.** 1965: *An introduction to statistical models in geology*. New York: McGraw-Hill.

Kruskal, J. B. 1964a: Multidimensional scaling by optimizing goodness of fit to a non metric hypothesis. *Psychometrika* 29, 1–27.

1964b: Non metric multidimensional scaling: a numerical method. *Psychometrika* 29, 115–29.

Lance, G. N. and **Williams, W. T.** 1965: Computer programs for monothetic classification ('association analysis'). *Computer Journal* 8, 246–9.

1966: Computer programs for hierarchical polythetic classification ('similarity analysis'). *Computer Journal* 9, 60.

1967: A general theory of classificatory sorting strategies: I. Hierarchical systems. *Computer Journal* 9, 373–80.

1968a: A general theory of classificatory sorting strategies: II. Clustering systems. *Computer Journal* 10, 271–6.

1968b: Note on a new information-statistic classificatory program. *Computer Journal* 11, 195.

Langley, R. 1968: *Practical Statistics*. London: Pan.

Lankford, P. M. 1968: *Regionalization: theory and alternative algorithms*. University of Chicago, Department of Geography: unpublished Master's dissertation.

1969: Regionalization: Theory and alternative algorithms. *Geographical Analysis* 1, 196–212.

Lawley, D. N. and **Maxwell, A. E.** 1963: *Factor analysis as a statistical method*. London: Butterworths.

Lockhart, W. R. and **Hartman, P. A.** 1963: Formation of monothetic groups in quantitative bacterial taxonomy. *Journal of Bacteriology* 85, 68–77.

Macka, M. editor 1967: *Economic regionalization.* Proceedings of the fourth general meeting of the Commission on Methods of Economic Regionalization of the International Geographical Union, Academia Publishing: Czechoslovak Academy of Sciences.

Mackay, J. R. 1958 : Chi square as a tool for regional studies. *Annals of the Association of American Geographers* 48, 164.

1959 : Comments on the use of chi square. *Annals of the Association of American Geographers*, 49, 89.

MacQueen, J. 1966 : Some methods for classification and analysis of multivariate observations. *University of California, Western Management Science Institute, Working Paper* 96.

Mahalanobis, P. C. 1936 : On the generalized distance measure in statistics. *Proceedings of the National Institute of Science, India* 2, 49–55.

Maisel, H. 1969: *A KWIC index to taxometric (numerical taxonomy) literature.* Georgetown University, Computation Centre.

Mannetje, L. T. 1967 : A comparison of eight numerical procedures applied to the classification of some African *Trifolium* taxa based on Rhizobium affinities. *Australian Journal of Botany* 15, 521–8.

Mayfield, R. C. 1967 : A central-place hierarchy in northern India. In Garrison, W. L. and Marble, D. F., editors, Quantitative geography, Part I : economic and cultural topics, *Northwestern University, Studies in Geography* 13, 120–66.

McElrath, D. and **Barkey, J. W.** 1964 : Social and physical space : models of metropolitan differentiation. Northwestern University, Centre for Metropolitan Studies : unpublished paper.

McNaughton-Smith, P. 1965 : Some statistical and other numerical techniques for classifying individuals. *Home Office Research Unit Report* 6. London : H.M.S.O.

McNaughton- Smith, P., Williams, W. T., Dale, M. B. and **Mockett, L. G.** 1964 : Dissimilarity analysis : a new technique of hierarchical subdivision. *Nature* 202, 1034–5.

McQuitty, L. L. 1957 : Elementary linkage analysis. *Educational and Psychological Measurement* 17, 207–29.

1960 : Hierarchical syndrome analysis. *Educational and Psychological Measurement* 20, 293–304.

1963 : Rank order typal analysis. *Educational and Psychological Measurement* 23, 55–61.

1966 : Single and multiple classification by reciprocal pairs and rank order types. *Educational and Psychological Measurement* 26, 253–65.

1968 : Multiple clusters, types and dimensions from iterative intercolumnar correlational analysis. *Multivariate Behavioral Research* 3, 465–78.

McQuitty, L. L. and **Clark, J. A.** 1968 : Clusters from iterative intercolumnar correlational analysis. *Educational and Psychological Measurement* 28, 211–38.

Megee, M. 1965 : Economic factors and economic regionalization in the United States. *Geografiska Annaler* 47B, 125–37.

Michener, C. D. and **Sokal, R. R.** 1957 : A quantitative approach to a problem in classification. *Evolution* 11, 130–62.

Miller, R. L. and **Kahn, J. S.** 1962 : *Statistical analysis in the geological sciences.* New York : Wiley.

Moser, C. A. and **Scott, W.** 1961 : *British towns: a statistical study of their social and economic differences.* Centre for Urban Studies, Report 2. Edinburgh : Oliver and Boyd.

Murdie, R. A. 1968 : The factorial ecology of Metropolitan Toronto, 1951–1961. An essay on the social geography of the city. *University of Chicago, Department of Geography, Research Paper* 116.

Needham, R. M. 1962 : A method for using computers in information classification. *Proceedings of IFIP Congress* 62, 284–6.

Neely, P. M. N.D.a: General discussion re : neighbourhood limited classification. University of Chicago : unpublished paper.

N.D.b : Toward a theory of classification. University of Chicago : unpublished paper.

1965: Grouping with contiguity constraint. UCSL 501.

Ng, R. 1969: Recent internal migration in Thailand. *Annals of the Association of American Geographers* 59, 710–30.

Norcliffe, G. 1968 : Areal grouping with elementary and nearest neighbour linkage analysis. *University of Bristol, Department of Geography, Seminar Paper, series A* 12.

Norman, P. 1968 : The classification of enumeration districts in the Third Survey of London. *IBG Study Group in Statistical Techniques, London, October, 1968.*

Nystuen, J. D. and **Dacey, M. F.** 1961 : A graph theory interpretation of nodal regions. *Regional Science Association, Papers and Proceedings* 7, 29–42.

Olson, E. C. and **Miller, R. L.** 1958 : *Morphological integration.* University of Chicago Press.

Orlocci, L. 1966 : Geometric models in ecology, I : The theory and application of some ordination methods. *Journal of Ecology* 54, 193–215.

1967 : An agglomerative method for classification of plant communities. *Journal of Ecology* 55, 193–206.

Pears, N. V. 1968 : Some recent trends in classification and description of vegetation. *Geografiska Annaler* 50A, 162–72.

Pearson, K. 1900 : On the criterion that a given system of deviations from the probable in the case of a correlated system of variables is such that it can be reasonably supposed to have arisen from random sampling. *Philosophical Magazine*, Series 5, 50, 157–72.

1926 : On the coefficient of racial likeness. *Biometrika* 18, 105–17.

Peters, P. 1968 : The urban ecology of Louisville, Kentucky. University of Chicago, Centre for Urban Studies : unpublished paper.

Peterson, G. L. 1967 : A model of preference : quantitative analysis of the perception of the visual appearance of residential neighbourhoods. *Journal of Regional Science* 7, 19–32.

Pocock, D. C. D. and **Wishart, D.** 1969 : Methods of deriving multi-factor uniform regions. *Transactions of the Institute of British Geographers* 47, 73–98.

Prakasa Rao, V. L. S. 1953 : Rational groupings of the districts of Madras State. *Indian Geographical Journal* 28, 33–43.

Pyle, G. F. 1968 : *Some examples of urban medical geography.* University of Chicago, Department of Geography : unpublished Master's dissertation.

Rao, C. R. 1948 : The utilization of multiple measurements in problems of biological classification. *Journal of the Royal Statistical Society* 10, 159–93.

Ray, D. M. and **Berry, B. J. L.** 1966 : Multivariate socioeconomic regionalization : a pilot study in Central Canada. In Ostry, S. and Rymes, T., editors, *Regional statistical studies,* University of Toronto Press, 75–130.

Rees, P. H. 1968 : *The factorial ecology oj metropolitan Chicago, 1960.* University of Chicago, Department of Geography : unpublished Master's thesis.

1969 : Factorial ecology : an extended definition, survey and critique of the field. *IGU Commission on Quantitative Methods, Ann Arbor, Michigan, August, 1969.*

In press : The factorial ecology of metropolitan Chicago (1960). In Berry, B. J. L. and Horton, F., editors, *Geographic perspectives on urban systems,* Englewood Cliffs, N.J. : Prentice Hall.

Rice, J. 1965 : Patterns of Swedish foreign trade in the late eighteenth century. *Geografiska Annaler* 47B, 86–99.

Rodgers, D. J. and **Tanimoto, T. T.** 1960 : A computer program for classifying plants. *Science* 132, 1115–18.

Rose, M. J. 1964 : Classification of a set of elements. *Computer Journal* 7, 208–211.

Rubin, J. 1967 : Optimal classification into groups : an approach for solving the taxonomy problem. *Journal of Theoretical Biology* 15, 103–44.

Rummel, R. J. 1967 : Understanding factor analysis. *Journal of Conflict Resolution* 11, 444–80.

Russett, B. 1966: Discovering voting groups in the United Nations. *American Political Science Review* 60, 327–39.

1967: Delineating international regions. In Singer, J. D., editor, *Quantitative International politics*. Glencoe: Free Press, 317–52.

Sawyer, W. W. 1966: *A path to modern mathematics*. London: Penguin.

Scott, A. J. 1968: Combinatorial processes, geographic space and planning. *University College, London, Department of Town Planning, Discussion Paper* 1.

1969a: On the optimal partitioning of spatially distributed point sets. In Scott, A. J., editor, *Studies in regional science*, London: Pion, 57–72.

1969b: A bibliography on combinatorial programming methods and their application in regional science and planning. *University of Toronto, Centre for Urban and Community Studies, Report* 1.

Shepard, R. N. 1962: Analysis of proximities: multidimensional scaling with an unknown distance function, I and II. *Psychometrika* 27, 125–140, 219–46.

Shepherd, M. J. and **Wilmott, A. J.** 1968: Cluster analysis on the Atlas computer. *Computer Journal* 11, 57–62.

Shevky, E. and **Bell, W.** 1955: Social area analysis: theory, illustrative application and computational procedures. *Stanford Sociological Series* 1, Stanford University Press.

Siegel, S. 1956: *Nonparametric statistics for the behavioral sciences*. New York: McGraw-Hill.

Silvestri, L., Turri, M., Hill, L. R. and **Gilardi, E.** 1962: A quantitative approach to the systematics of actinomycetes based on overall similarity. In Ainsworth, G. C. and Sneath, P. H. A., editors, *Microbial classification*, Cambridge University Press, 333–60.

Simpson, G. G. 1951: The species concept. *Evolution* 5, 285–98.

1961: *Principles of animal taxonomy*. New York: Columbia University Press.

Sinnhuber, K. A. 1954: Central Europe—Mitteleuropa—Europe Central: an analysis of a geographical term. *Transactions of the Institute of British Geographers* 20, 15–39.

Sneath, P. H. A. 1957: The application of computers to taxonomy. *Journal of General Microbiology* 17, 201–6.

1961: Recent developments in theoretical and quantitative taxonomy. *Systematic Zoology* 10, 118–39.

1962: The construction of taxonomic groups. In Ainsworth, G. C. and Sneath, P. H. A., editors, *Microbial classification*, Cambridge University Press, 289–332.

Snedecor, G. W. and **Cochran, W. G.** 1967: *Statistical methods*. Iowa State University Press. (Sixth edition.)

Sokal, R. R. 1961: Distance as a measure of taxonomic similarity. *Systematic Zoology* 10, 70–9.

Sokal, R. R. 1965: Statistical methods in systematics. *Biological Reviews* 40, 337–91.

Sokal, R. R. and **Michener, C. D.** 1958: A statistical method for evaluating systematic relationships. *University of Kansas Science Bulletin* 38, 1409–38.

Sokal, R. R. and **Rohlf, F. J.** 1962: The comparison of dendrograms by objective methods. *Taxonomy* 11, 33–40.

Sokal, R. R. and **Sneath, P. H. A.** 1963: *Principles of numerical taxonomy.* San Francisco: Freeman.

Sorensen, T. 1948: A method of establishing groups of equal amplitude in plants sociology based on similarity of species content and its application to analyses of the vegetation on Danish commons. *Biologiske Skrifter* 5, 1–34.

Spearman, C. 1913: Correlations of sums and differences. *British Journal of Psychology* 5, 417–26.

Spence, N. A. 1967: Multifactor uniform regionalization: a selected bibliographic guide. *London School of Economics, Graduate School of Geography, Discussion Paper* 6.

1968: A multifactor regionalization of British counties on the basis of employment data for 1961. *Regional Studies* 2, 87–104.

1969a: Classificatory analysis by discriminant iterations of English counties on the basis of employment data for 1961. London School of Economics, Department of Geography: unpublished paper.

1969b: Cross sectional analysis of English regional employment dimensions: 1951 and 1961. London School of Economics, Department of Geography: unpublished paper.

Spodek, H. 1968: The urban ecology of Shreveport, Louisiana. University of Chicago, Centre for Urban Studies: unpublished paper.

Stevens, B. H. and **Brackett, C. A.** 1968: Regionalization of Pennsylvania counties for development planning. In Berry, B. J. L. and Wróbel, A., editors, 154–87.

Stone, R. 1960: A comparison of the economic structure of regions based on the concept of 'distance'. *Journal of Regional Science* 2, 1–21.

Streumann, C. 1967: *Economic regionalization: a bibliography of publications of the German language.* Bad Godesberg: Bundesanstalt für Landeskunde und Raumforschung.

Sweetser, F. L. 1965a: Factorial ecology: Helsinki, 1960. *Demography* 1, 372–86.

1965b: Factor structure as ecological structure in Helsinki and Boston. *Acta Sociologica* 8, 205–25.

Taylor, P. J. 1968: Grouping analysis in regional taxonomy. *University of Newcastle upon Tyne, Department of Geography, Seminar Paper* 1.

1969: The location variable in taxonomy. *Geographical Analysis* 1, 181–95.

Taylor, P. J. and **Spence, N. A.** 1969: On contiguity in regionalization. University of Newcastle and London School of Economics, Departments of Geography: unpublished paper.

Thompson, D. 1967: Some comments on the relevance of multivariate analysis to geography. *Third Anglo-Polish Geographical Seminar, Baranow, Poland, September, 1967.*

Thompson, D. and **Hall, R.** 1969: A bibliographic guide to factor analysis methods and applications. University of Maryland, Department of Geography: unpublished paper.

Thompson, J. H., Sufrin, S. C., Gould, P. R. and **Buck, M. A.** 1962: Towards a geography of economic health: the case of New York state. *Annals of the Association of American Geographers* 52, 1–21.

Thorndike, R. L. 1953: Who belongs in the family? *Psychometrika* 18, 267–76.

Tryon, R. C. 1939: *Cluster analysis.* Ann Arbor: Edwards.

1955: Identification of social areas by cluster analysis. *University of California, Publications in Psychology* 8.

Tucker, L. R. 1963: Implications of factor analysis of three way matrices for measurement of change. In Harris, C. W., editor, *Problems of measuring change*, Madison, Wisconsin, 122–37.

1964: The extension of factor analysis to three dimensional matrices. In Frederikson, N. and Gullikson, H., editors, *Contributions to mathematical psychology*, New York: Holt, Rinehart and Winston, 110–27.

Van Arsdol, M., Camilleir, S. F. and **Schmid, C. F.** 1958: The generality of urban social area indices. *American Sociological Review* 23, 277–84.

Wallace, C. S. and **Boulton, D. M.** 1968: An information measure for classification. *Computer Journal* 11, 185–94.

Ward, J. H. Jr. 1963: Hierarchical grouping to optimise an objective function. *Journal of the American Statistical Association* 58, 236–44.

Ward, J. H. and **Hook, M.** 1963: Application of an hierarchical grouping procedure to a problem of grouping profiles. *Educational and Psychological Measurement* 23, 69–81.

Warntz, W. 1968: Some elementary and literal notions about geographical regionalization and extended Venn diagrams. In The philosophy of maps, *Michigan Inter-University Community of Mathematical Geographers, Discussion Paper* 12, 7–30.

Wayne, L. G. 1967: Selection of characters for an Adansonian analysis of mycobacterial taxonomy. *Journal of Bacteriology* 93, 1382–91.

Weaver, J. B. and **Hess, S. W.** 1963: A procedure for non partisan districting: development of computer techniques. *Yale Law Journal* 73, 288–308.

Whittlesey, D. 1954: The regional concept and the regional method. In James, P. and Jones, C. F., editors, *American geography: inventory and prospect*, Syracuse, 19–68.

Williams, W. T. and **Dale, M. B.** 1965: Fundamental problems in numerical taxonomy. *Advances in Botanical Research* 2, 35–68.

Williams, W. T. and **Lambert, J. M.** 1959: Multivariate methods in plant ecology, I: Association analysis in plant communities. *Journal of Ecology* 47, 83–101.

1960: Multivariate methods in plant ecology, II: The use of an electronic digital computer for association analysis. *Journal of Ecology* 48, 689–710.

Wishart, D. 1968: *A Fortran II program for numerical classification.* St Andrews.

1969a: Numerical classification method for deriving natural classes. *Nature* 221, 97–8.

1969b: The use of cluster analysis in the classification of diseases. *Proceedings of the Scottish Society of Experimental Medicine*, *Scottish Medical Journal* 14, 96.

1969c: An algorithm for hierarchical classifications. *Biometrics* 22, 165–70.

1969d: Fortram II programs for 8 methods of cluster analysis (Clustan I). *State Geological Survey*, *University of Kansas*, *Computer Contribution* 38.

Wrigley, E. A. 1965: Changes in the philosophy of geography. In Chorley, R. J. and Haggett, P., editors, *Frontiers in geographical teaching*, London: Methuen, 3–20.

Zobler, L. 1957: Statistical testing of regional boundaries. *Annals of the Association of American Geographers* 47, 83–95.

1958a: Decision making in regional construction. *Annals of the Association of American Geographers* 48, 140–8.

1958b: The distinction between relative and absolute frequencies in using chi-square for regional analysis. *Annals of the Association of American Geographers* 48, 456–7.

Geographic space perception

past approaches and future prospects

---

by Roger M. Downs

Contents

I Introduction

IN his review of the quantitative revolution in geography, Burton (1963, 158) referred to 'the problem of perception, which may soon come to merit a place alongside the quantitative revolution in terms of significant new viewpoints.' Considering the current impetus and volume of research into geographic space perception, sufficient time has elapsed to enable one to see if Burton's great expectations have been realised. Even the most fervent proponent of the current view (that human spatial behaviour patterns can be partially explained by a study of perception) would admit that the resultant investigations have not *yet* made a significant contribution to the development of geographic theory.

This paper comprises three sections. The first will commence with a discussion of the objectives involved in the study of geographic space perception. This will be followed by a review of some of the recent approaches to the problem of perception in geography. This section will conclude with some comments on issues which are crucial for future research into perception. The second section will develop a conceptual schema or strategy which will allow future research to be explicitly formulated and developed in close conjunction with relevant research in other social sciences. It will concentrate on the contribution that other disciplines can make to an understanding of our particular spatial problems. The third section will discuss the tactical problems inherent in translating concepts into specific research topics. It will focus upon the problem of measurement which is fundamental to the development of research in perception. Some of the technical methods currently available for solving these problems will be reviewed.

II Recent approaches to the problem of geographic space perception

1 Objectives in the study of geographic space perception

When forced to give a definition of their subject, many geographers suggest that they are studying the man/land ecosystem, or that they are concerned with the earth as the home of man. The common theme of these, and the many similar definitions, is the expression of the idea of man/land relationships: these are essentially ecological relationships. The Sprouts (1965) have traced the academic pedigree of this ecological view of our subject back through the families of possibilism and probabilism to the controversial idea of environmental determinism. Indeed, Rostlund (1956, 23) argues that 'environmentalism was not disproved; only disapproved', and then demonstrates that environmentalism is a concept which still pervades much of our thinking. How does the current interest in geographic space perception fit into this background, and what is the objective of such research?

The study of geographic space perception is only part of a more general trend in modern geography, namely the 'behavioural revolution' which seems to be closely following in the wake of the quantitative revolution. We can explain the behavioural revolution by using an analogy drawn from communications theory. The behavioural revolution represents a fundamental change in our conceptual approach to understanding human spatial behaviour, and is characterised by a more realistic view of man, in combination with the use of quantitative methods. Prior to this, man was regarded as a 'black box' concept; that is, an unknown constant in the study of the environment/behaviour relationship. The tendency was to make certain assumptions about man which then made it possible to understand spatial behaviour primarily in terms of the environment. The behavioural approach replaces the black box concept of man by a 'white box'. Thus more realistic assumptions about the nature of man, drawn largely from other social sciences, are employed, and mean that the basic schema for analysis is no longer environment/spatial behaviour, but environment/man/spatial behaviour. Man therefore becomes an intervening variable, and in this behavioural formulation is a significant, if not crucial, variable.

We can illustrate the development of the behavioural approach by considering central place theory. In its original formulation, the theory involves the simplifying assumption that man is an 'economic man': that is, he is completely informed, infinitely sensitive, and rational (Edwards, 1954, 381). That this is an incorrect assumption is obvious. But it is not the immediate intuitive implausibility of this assumption that we must question. Instead, we must concern ourselves with its consequences when we come to compare predicted or expected behaviour patterns with actual behaviour patterns. For example, Marble (1959) found that in Detroit, Washington State, Jefferson City and Cedar Rapids, the total distance travelled and the trip frequency were not significantly affected by the relative location of the individual. Yet given the location of an individual relative to the retail facilities, and the spatial distribution of types of retail function, such travel behaviour should display predictable regularities in terms of the theory. Marble did find, however, that consumer travel behaviour was significantly related to the social and economic characteristics of the individual.

Such findings in the context of human spatial behaviour are not uncommon, and there are two solutions to the problems that they present to the development of theory. Given that the original assumptions are still reasonable, we can consider the predicted relationships between the variables not to be deterministic but probabilistic. Harvey (1966, 82) suggests that we can view behavioural constraints:

> . . . as a random disturbance effect upon the ideal patterns generated by classical theory. A probability test of a normative theory is quite feasible. The question then becomes one of identifying when the real world situation deviates from the ideal by more than some 'random' component, where the random component summarizes the imperfections inherent in the decision-making process.

Or, and this is the basis of the behavioural approach, we can replace these original assumptions by ones which are more realistic and adequate expressions of man's nature. Thus the variable nature of man's capabilities and limitations is allowed to intervene between the environment and the spatial behaviour pattern in our attempts at explanation and theory development. The behaviour pattern, which can be explained in terms of a numerical prediction by a stochastic formulation, is subject to greater analysis and therefore understanding by adopting the environment/man/ spatial behaviour schema.

To return to the example of consumer spatial behaviour, the limiting assumptions involved in the economic man concept can be replaced by assumptions more in accordance with our present knowledge of man's nature. Golledge (1967, 239) in his approach to marketing behaviour, views the problem in a learning theory framework and relaxes the concept of perfect knowledge:

> From an initial evaluation of the structure of a tributary area, some behavioral characteristics associated with different sectors of tributary areas are isolated. These characteristics are interpreted as representing stages in a learning process, and a relation is drawn between learning theory and market area analysis.

Similarly, Huff (1959; 1960) suggests that spatial behaviour is dependent upon the individual's evaluation and relative assessment of the environment: in this instance the environment consists of the retail facilities and the transport routes and media linking these to the individual. By introducing the twin concepts of 'consumer space perception' and 'consumer space preference', Huff simultaneously relaxes the concepts of infinite sensitivity and rationality in the economic man assumption. Given that both consumer space perception and preference are functionally related to the social and economic characteristics of the individual, Huff argues that this offers an explanation of Marble's findings. In this way the original theory is made capable of greater predictive power, and the section of the real world incorporated in it is better understood.

The behavioural approach represents an attempt to employ more realistic concepts of man in the analysis of human spatial behaviour. Thus the black box is replaced by a white box, in which the variable nature of man is recognised as being of fundamental importance. Perception studies concentrate on the cognitive understanding that man has of his environment and the way in which this knowledge is stored and organised in the mind: that is, they are concerned with the image of the real world. One of the principal underpinnings of the perception approach is that spatial behaviour is a function of the image, where the image represents man's link with his environment. The focus is on such problems as the nature of the image, its relationship to the real world, the processes which influence the transformation of the real world into the image, and the precise effect of images on decisions. In addition these problems have a dynamic aspect through time, and vary across different groups of people. Obviously, therefore, the problem of geographic space perception is multifaceted and highly complex. Yet it represents a significant advance in our conceptual approach to spatial behaviour since it throws more emphasis on the underlying processes rather than on the spatial patterns that behaviour produces.

Studies of geographic space perception form a logical step in the attempts to understand man/environment relationships. They represent a point somewhere between possibilism and determinism, but are conceptually and technically more sophisticated than probabilism. Our task is to focus on the image and to do this by employing the accumulated wisdom of the social sciences in general, and our newly acquired technical skills and competences in particular. I shall now review some of the recent approaches to the problem of perception in geography.

2 The types of research

Three major approaches can be distinguished and these will be reviewed in sequence. They are not mutually exclusive, but represent a convenient typology for discussing the published work on geographic space perception.

a The Structural approach: The first, typified by the work of Appleyard and others (1964), Heinemeyer (1967), Lee (1964a and b), Lucas (1964), Lynch (1960), Steinitz (1967) and Trowbridge (1913), is concerned with the identity and structure of geographic space perceptions. It has an interesting parallel in the general approach to geography's fundamental problem, the nature of spatial division. It is one of the few coherent features of geography that it has proved necessary and desirable to divide space into some form of regional system. Ignoring the controversy over methods of division, we recognise that the need for division stems from a desire to order our knowledge of the world. Similarly, there must be some mental basis for the storage of relevant information from the environment. We are therefore working on the assumption of a limited or finite capacity for such information storage. Given this, we can ask several basic questions, all of which are centred on the concept of the image. What information is stored in our minds? How is it stored: that is, how is it structured? What is the relationship between the image (or stored information) and the real world from which the image is abstracted? What is the process of abstraction determined by?

There is a basic theme distinguishable in the work on the mental or cognitive organisation of space. Namely, man must *orientate* himself successfully in order to be able to implement decisions as behaviour. Thus, to do something we must know where to do it, and we must relate ourselves to certain abstracted features of the environment. The earliest approach to the problem of orientation was Trowbridge's (1913, 888) 'study of the reasons why civilised man is so apt to lose his bearings in unfamiliar regions.' His most relevant ideas concern imaginary maps, which he defined as 'the conception of a mental image of an orientation map that is entirely imaginary, and erroneous' (1913, 890). On the basis of empirical research, he distinguished seven types of imaginary map, and suggested that the cause of such maps is loose early education and erroneous later impressions.

Trowbridge's work is, or should have been, a model for later research. He set up hypotheses, tested them using data, interpreted the results, and then suggested the type of processes that may have been operating. However the theme was not taken up again until Lynch's work in 1960. (For a review of psychological work in this area, see Howard and Templeton, 1966, chapter X). Lynch was concerned with orientation, deriving many of his basic ideas from psychology. He conducted extensive field studies in three cities, Boston, New Jersey and Los Angeles, concentrating on the images that inhabitants held of these cities. The type of image produced is illustrated in the sequence of Figures 1–6, all taken from his studies of Boston. The most important products of his work are the twin concepts of legibility and imageability, together with the five element types which form the structural bases of the mental images. These element types comprise paths, edges, districts, nodes and landmarks. They are the components out of which the inhabitants synthesise their images: as such, they are clear expressions of the relationship between the image of the city and the ability to live in it (i.e. to express decisions as overt behaviour within the urban environment). Figure 1 represents an objective

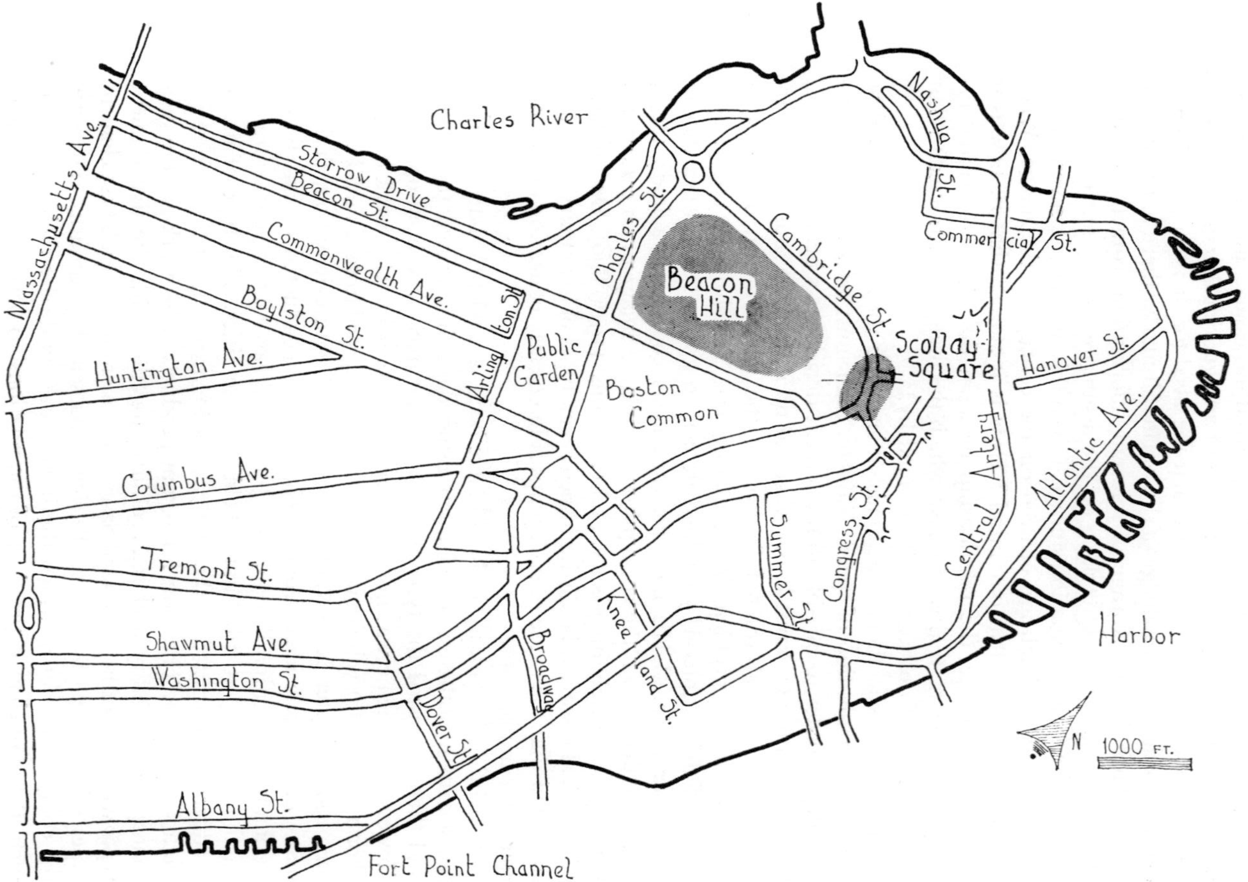

Fig. 1. Outline map of the Boston Peninsula. *From Lynch* (1960)

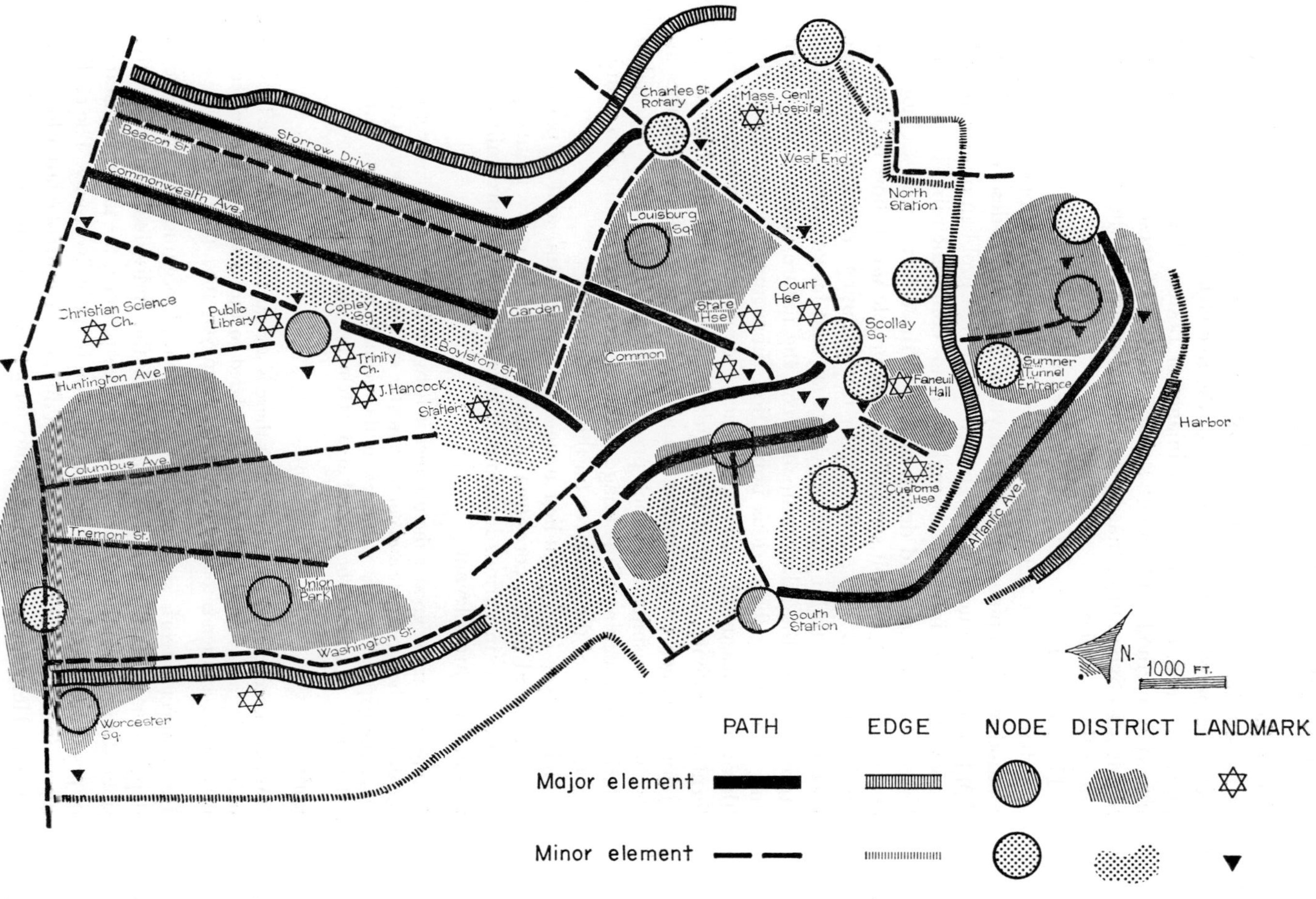

Fig. 2 The visual form of Boston as seen in the field. *From Lynch* (1960)

datum (i.e. a conventional map); Figure 2 shows the major visual element of Boston derived by trained field observers. The latter and succeeding maps are based on a combination of the five element types. Figures 4–7 indicate the differing nature of images of Boston, derived by different interview methods. The most striking feature of these images is their highly selective nature: that is, the mental space and the objective space are distinct entities.

The power and generality of Lynch's analytic approach to the problem of urban images is demonstrated by the extensive application of the method in urban planning in recent years. Two studies deserve comment. Appleyard and others (1964) studied the design problem of urban roadscapes, with the intention of making cities comprehensible by manipulation of the pattern and design of roads: 'the driver would see how the city is organised, what it symbolises, how people use it, how it relates to him' (Appleyard *et al.*, 1964, 2). In this planning situation, the image becomes the controlling link between man and his environment. Steinitz (1967) studied the relationship between the nature of the environment and its meaning for various population subgroups. He postulated that 'there are measurable correspondences—congruences—between urban form and activity, and that the regularities in these relationships have a major influence on the amounts and kinds of meanings which the environment transmits and which people acquire.' This work is unusual in relating the meaning of the image to the environment, whilst also examining the effect of the perceiver on this relationship.

Heinemeyer (1967) studied images of Amsterdam, and in particular was interested in the core area of the city and its entry points as seen by individuals in various locations. These entry points are called gates. two findings are relevant:

> It appears that the greater the distance is between home and core, the greater is the distance between gate and core. To say the same thing in a somewhat different way, the expanding city witnesses in the mental experience of its citizens an enlargement of scale of the central area. (Heinemeyer, 1967, 87)

and:

> It might be concluded tentatively that there is a clear tendency that the location of inter-city activities (i.e. shopping) is related to the location of the subjective urban core, and tends to vary with it. (Heinemeyer, 1967, 98)

Heinemeyer's work suggests that the mental organisation of space is dependent upon the relative location of the individual: consequently, since the image is the basis of the behaviour pattern, the factor of distance or proximity re-enters into the formulation in an indirect way. The earlier discussion of central place theory dealt directly in terms of relative location, whereas Heinemeyer's finding allows for a further modification in accordance with Huff's concept of consumer space perception. Lucas (1964) also demonstrated that the spatial extent of an image depended upon the intended behaviour pattern of an individual. He found that the perceived size of a recreation area depended upon the type of user—for example, the area as viewed by a motor canoeist was different from that as seen by a paddle canoeist. This suggests that the relationships between the nature of the image, the relative location of the individual, and their behaviour will be extremely complex.

Lee (1964a and b) was concerned with the urban planning concept of a neighbourhood and its mental or cognitive correlate. His objective was to see if people's behaviour patterns and knowledge were organised in a way compatible with current planning concepts. Was the neighbourhood based on an aggregate of people or an area of ground? He found that these were inextricably linked and that people could draw on a map a line around the area which served as their neighbourhood. He argued that 'repeated transaction with people and places in the urban environment leads, by a process of differentiation, to the separation of an organised socio-spatial whole' (1964a, 7). His approach to the problem of the nature of the image is interesting because he related the concept of an image to an existing psychological concept, a schema which can be considered as the mental organisation pattern that determines a behavioural process. Thus Lee made a direct attempt to link a current planning concept (i.e. a neighbourhood) and existing theory in one of the social sciences.

How can we draw together the findings of this approach? There appear to be three interesting issues for future research. First, in what form is spatial information stored? When we use the term 'mental map' (Gould, 1966), we imply that people preserve the relative location qualities of environmental data as in a conventional map. However, although the map is a good model for geographers, it does not follow that it is equally valid as a model for the mental 'ordering' of environmental data.

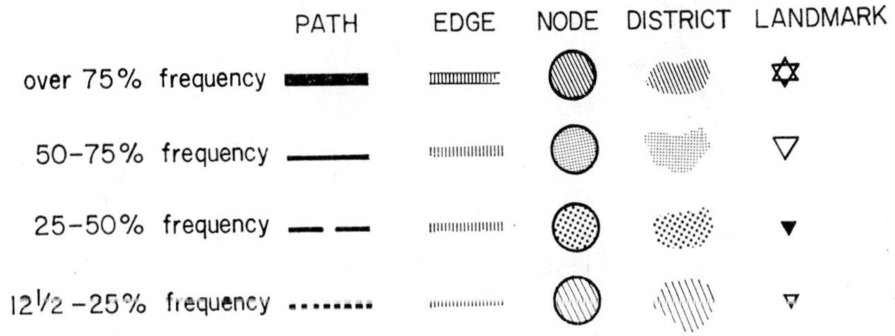

PATH EDGE NODE DISTRICT LANDMARK

over 75% frequency

50–75% frequency

25–50% frequency

12½–25% frequency

Fig. 3 Key to Figures 4–6. *From Lynch* (1960)

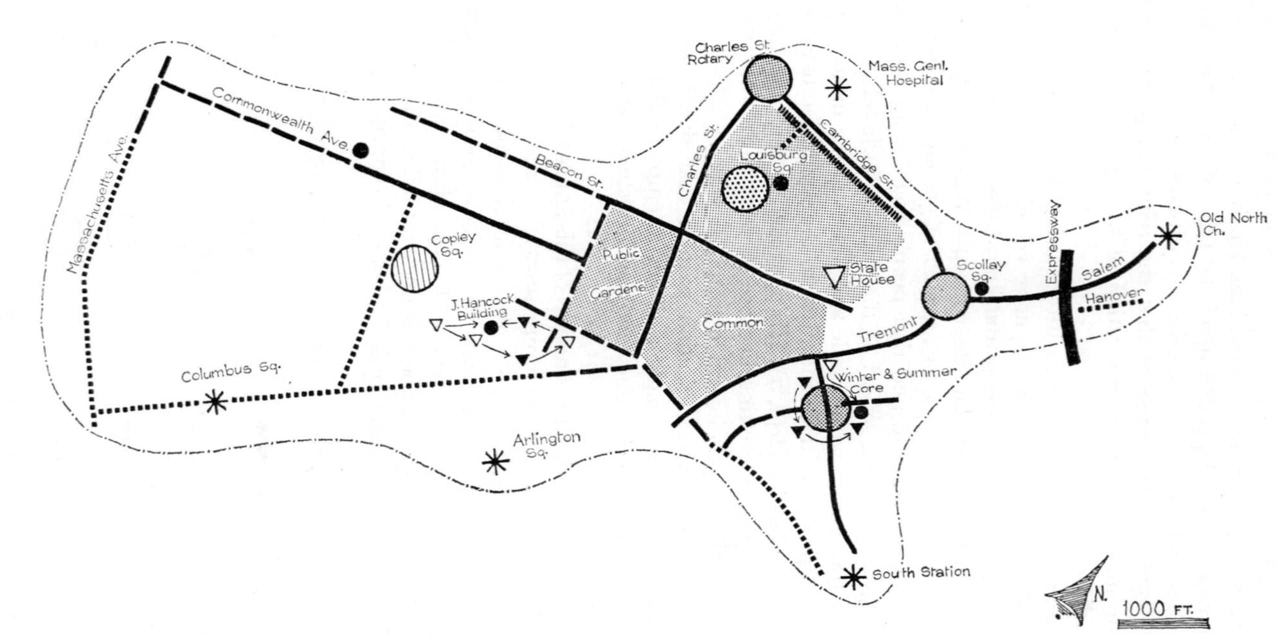

Fig. 4 The Boston image derived from street interviews. *From Lynch* (1960)

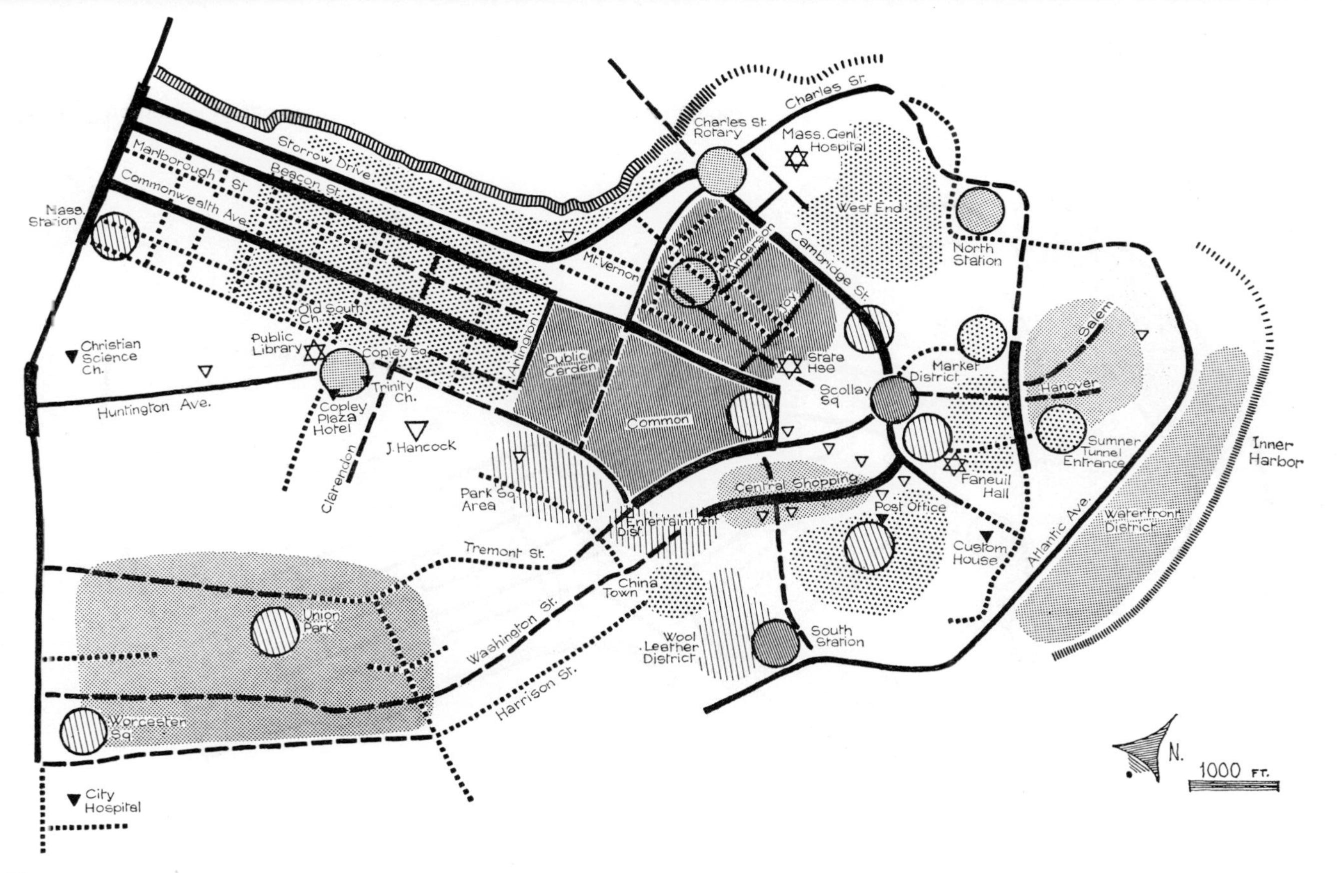

Fig. 5 The Boston image derived from verbal interviews. *From Lynch* (1960)

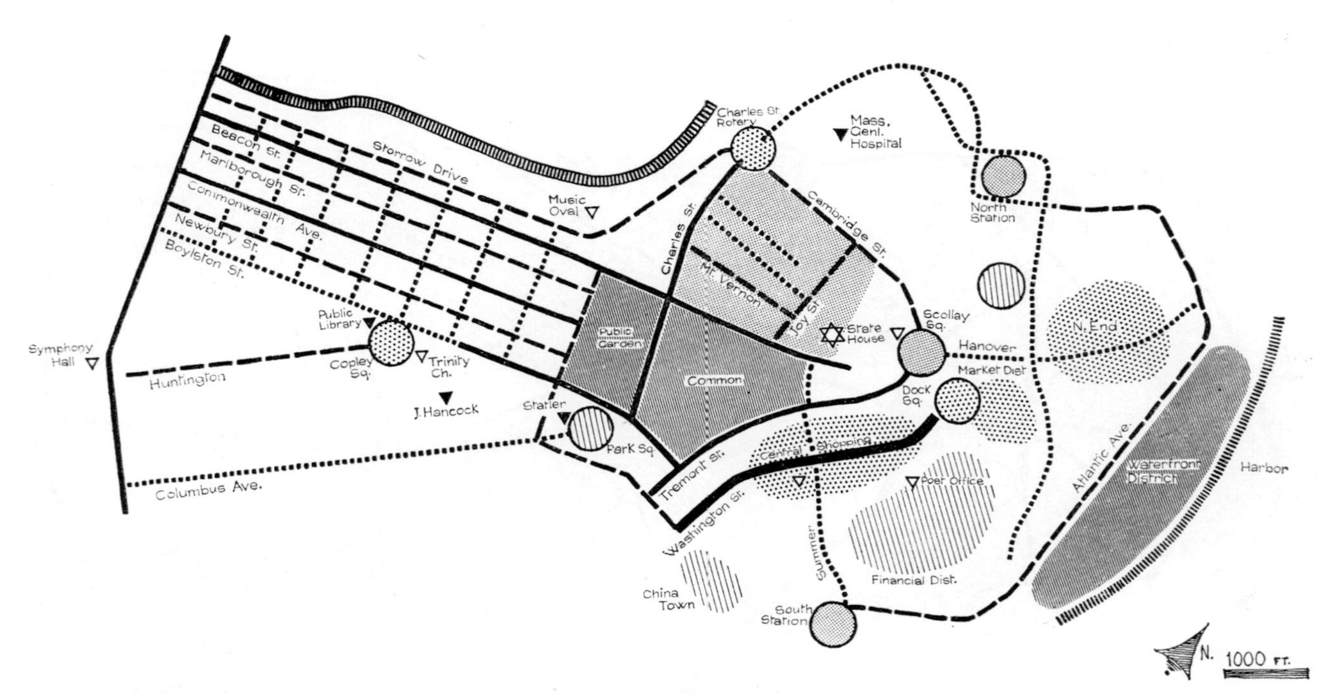

Fig. 6 The Boston image derived from sketch maps. *From Lynch* (1960)

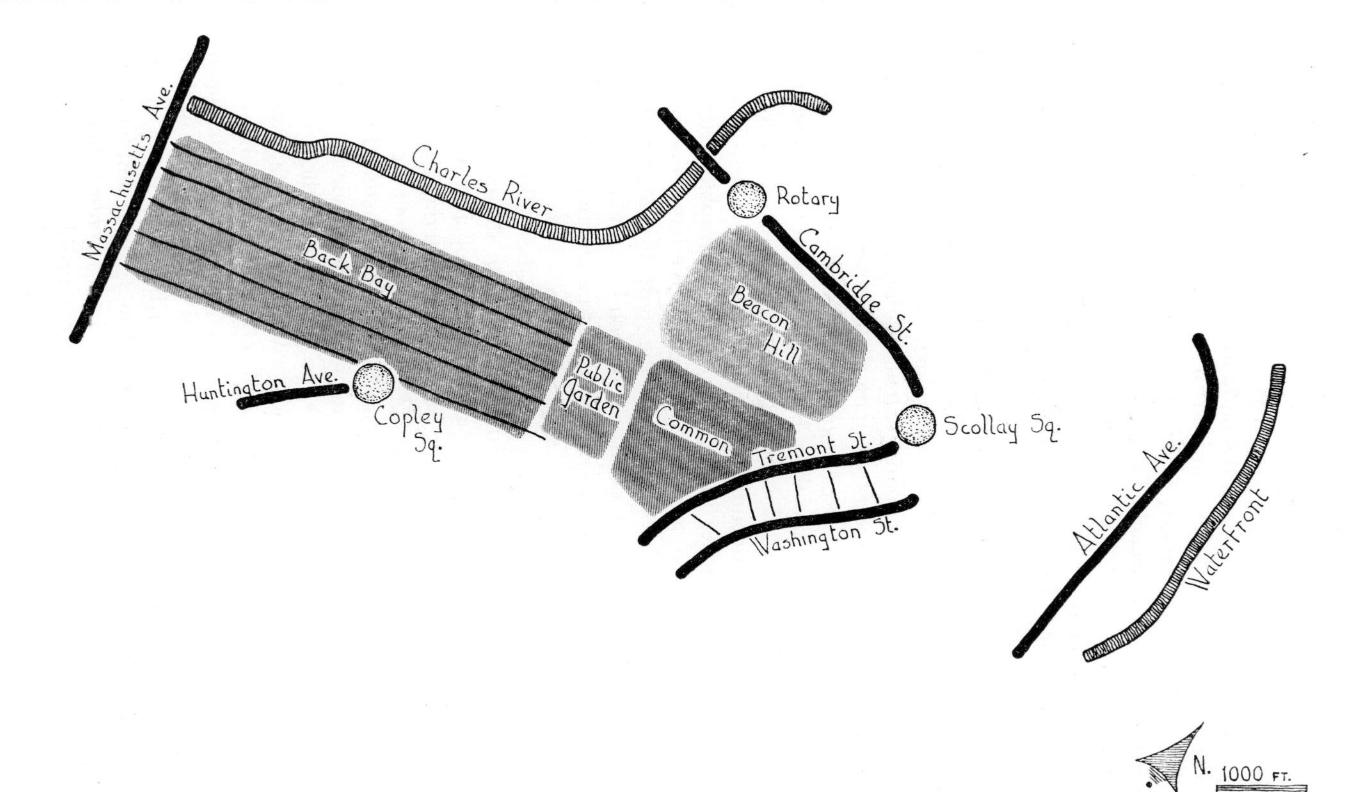

Fig. 7 The Boston that everyone knows. *From Lynch* (1960)

Second, is mental space continuous areally? Lee (1964a and b) assumed that it was, and that the enclosed line on a map delimited a complete neighbourhood. From a study of Lynch's maps, one could argue, however, that spatial images are bitty and discontinuous. It seems a logical extension to argue that mental space consists of a series of networks, with nodes representing important points, and links forming routes. Thus mental space may be better approached using topology than any other form of geometry.

Third, the findings of Lee (1964a and b) and Heinemeyer (1967) suggest that there may be a hierarchy of different images according to the intended space use, and that the image of the environment therefore depends upon the particular behaviour. Lucas (1964) demonstrated that the delimitation of the spatial extent of a recreation area depended upon the behavioural intentions of the individual. Similarly Webber (1964) organised his discussion of the spatial extent and interaction of individual roles in terms of a hierarchy centred upon each individual. Each of these roles must carry with them some cognitive organisation of the space in which they are played out. But there is no indication of how a hierarchy of spatial images is arranged. For example, are the images distinct or are they variations based on a common skeletal framework?

The other two approaches to the problem of geographic space perception are the 'evaluative' and the 'preference'. We have concentrated on the structural approach because it is the most disjointed and lacks an adequate summary. The evaluative approach is more familiar, whilst the preference approach is at present confined to a few researchers.

b The evaluative approach: This is concerned with the evaluation of the environment via spatial images, and seeks to relate the evaluation to decision making and therefore to behaviour. An implicit assumption is that the perceived world is one of the fundamental criteria or bases used in making a decision, which is then expressed as behaviour (see Gould's (1966) discussion of this concept). The question is, what factors do people consider important about their environment, and how, having estimated the relative importance of these factors, do they employ them in their decision making activities. The approach seeks to go one stage further than the previous one in that the structural components of the image are assigned weightings? In terms of decision theory, this focus on the image is the utility of various environmental states and the probability of their occurrence. The principal work has been contributed by the 'Chicago School', including White (1945), Burton and Kates (1964) and Saarinen (1966). Under the heading of hazard perception, a central theme has been the measurement of the probabilities that people attach to the occurrence of potentially dangerous environmental phenomena. This has led to a heavy concentration of interest in flood plain and coastal areas. Kates (1962), for example studied people living in areas subjected to sporadic riverine flooding. Given that each person had some information available, and therefore had formed some estimate of the flood potential, how was their pattern of behaviour affected? For a comprehensive review of this work, see Burton and Kates (1964) and Saarinen (1966).

c The preference approach: The third approach is based on the concept of preference. The basic question is, given a set of spatially differentiated objects, how do people assess these on a scale of preference with relation to some specified behaviour objective? The principal contributions are those of Gould (1966; 1967) and White (1967). Methodologically, all of these are similar, employing a factor analytic model on sets of rank orderings of objects (in the case of Gould, these objects were states in the United States, and countries in Europe and West Africa). The intention is to infer the underlying structure, and hence causes, of the spatial patterns of residential preference. The approach can be viewed as an indirect method of deriving the answers obtained by adopting the evaluative approach. Wolpert (1965) used the concept of preference to construct a model of migration based upon the unequal attractiveness of space, in which migration is a function of the attractiveness of a place, the perception of it, and the nature of the perceiver.

The major conclusion to be drawn from this over-view of recent work in geographic space perception is that, as a field of research, it is still in its infancy. If we may extend the analogy, it is also true to say that the early years of life are fundamental in determining the future development of our work. Olsson (1967, 6) argues that 'the vigorous group engaged in this sort of research must now prove the practical usefulness of their seemingly good ideas by translating them from often cumbersome English into simplified but operational and testable models'. It seems that we have reached a point where we must pause to consider the future. Olsson is arguing that the research should be directed towards the achievement of a set of goals that are part of the philosophy and objectives of the quantitative revolution. In particular, this involves the development of models and theories: a measure of the success of such objectives can be seen in the volumes on spatial analysis by Berry and Marble (1967) and locational analysis by Haggett (1965). Given that this philosophy is reasonable, there appear to be three areas that we must pay special attention to in the study of geographic space perception. We will turn to these next.

3 Problem areas for future research

There are three issues which seem important for the development of research in perception. Given that the development of theory concerning the spatial patterns and processes of human behaviour is one of our objectives, we must focus our attention on these issues.

The first is the necessity for an approach to the problem of perception via the development and testing of explicit hypotheses. Although exploratory studies serve a vital part in the early development of a discipline, we have advanced far enough to make hypothesis formulation and testing both necessary and desirable. The studies of drought hazard perception by Saarinen (1966) and of residential preference by Peterson (1967) both indicate the value of a methodological approach which tests explicit hypotheses.

A second issue concerns the status of our research with respect to the other social sciences. We should adopt an attitude which permits us to learn readily from the social sciences since our field of concern is essentially an inter-disciplinary one. The *Directory of Behaviour and Environmental Design* (1967) indicates the range of disciplines which have a basic interest in the problem of man/environment relationships. Kaplan (1964, 4) argues that 'in the world of ideas there are no barriers to trade or travel. Each discipline may take from others techniques, concepts, laws, models, theories, or explanations—in short, whatever it finds useful in its own inquiries.' However Sonnenfeld (1967, 53) has suggested that we, as human geographers, may have over-reacted from the effects of the criticism of environmental determinism:

> . . . the anti-environmentalist legacy of that era has discouraged the study of man's environmental sensitivities and behaviours. Without question, this has been a loss to geography; it may well have been a loss to behavioural science as well.

We must ensure that we re-establish our position in the behavioural or social sciences, and employ the accumulated wisdom that they have developed. In particular, we must borrow concepts, theories and techniques wherever they seem applicable. Craik (1968, 34) has remarked that:

> If the subtlety of reactions to the everyday physical surroundings has been one factor in hampering the development of behavioural science research in this area, as it indeed appears to have been, then the psychologist, if anyone, should be able to make a contribution. Responses are, after all, the business of the psychologist.

However, in order to communicate with the other social sciences, we must know and clearly state our objectives in the study of geographic space perception. In section II(1) I have tried to outline our reasons for adopting an approach to spatial behaviour which incorporates the idea of perception. We must also develop a conceptual schema which will allow us to state our objectives in broad terms. This will enable us in turn to borrow from the other social sciences in a systematic, productive way. There have been several previous attempts to provide a framework for this purpose—see Kirk (1951; 1963), Lowenthal (1961) and the Sprouts (1965). Of these, Lowenthal's work has received the greatest recognition: its major asset is the inclusion of a wide range of material from psychology.

The objective of the conceptual schema, which will be developed in section III, is to allow research in geographic space perception to be developed in close co-operation with research in the spectrum formed by psychology, social psychology and sociology.

III A conceptual schema for the study of geographic space perception

1 A justification of the need for a schema

The concept of a research frontier has become popular in geography. Extending the analogy further, most frontier movements involve some idea of direction or aim. Whether this direction is a misconception or not, it is crucial that it should be explicitly stated. Such a direction is notably lacking in our research into the relationships between perception and behaviour.

Lewin (1963, chapter 1) devotes a whole article to warning psychologists about the dangers of misdirected research. Drawing an analogy between scientific exploration and road-building on an unexplored continent, he argues that there is a danger of building academic super-highways that eventually lead to nowhere. Although we can never be certain that such a fate will not befall our research, we can reduce the chances of this happening by making clear, both to ourselves and to others, just where we do intend to go. A conceptual schema should be designed to fulfil this role.

The schema that we will develop also fits into the idea of the cyclical stages of scientific research discussed by Deutsch (1966). He identifies two stages. First, the philosophic stages which 'are concerned with strategy; they select the targets and the main lines of attack'; second, the empirical stages which 'are concerned with tactics; they attain the targets, or they accumulate experience indicating that the targets cannot be taken in this manner and that the underlying strategy was wrong.' He continues his argument thus:

> The social sciences today perhaps are approaching another 'philosophic crisis'—an age of re-examination of concepts, methods, and interests, of search for new symbolic models and/or new strategies in selecting their major targets for attack. (Deutsch, 1966, 3–4)

The recent history of human geography clearly demonstrates that the discipline has reached a philosophic crisis. Some progress has been made in the case of the rapid development of regional science and in the introduction of systems theory into spatial location theory. The field of geographic space perception is in dire need of a philosophic stage which can, by using the experience of its own past and that of other cognate disciplines, produce a coherent conceptual schema to direct and guide future research. The following schema is intended to provide a research strategy serving this purpose. However, such an objective is *not* in itself sufficient. Not only must it point out the directions of future research, but it must also be feasible—that is, capable of being operationalised. Therefore, section IV of this paper is intended to investigate the tactical problems of converting any conceptual schema for geographic space perception into testable hypotheses which will lead to the development of spatial theory.

2 The conceptual schema

A diagrammatic version of the schema is shown in Figure 8. The boxes represent concepts and the directional arrows indicate the links or relations between these concepts. Before discussing the schema in detail, two fundamental points must be understood. First, the schema is a blatant over-simplification of a highly complex situation. Certain concepts are extremely primitive and simplistic, whilst the directional linkages could be restructured until every concept had a direct link with every other one. Second, it is not an operational model: it will not lead directly to testable hypotheses. (For a discussion of this problem see section IV(1).)

Having stressed the flaws of the schema, what is its main value? The principal intention is to make explicit certain relationships which have been discussed in various studies of geographic space perception. In particular, these relationships have been structured so that they fit the current conceptions and theories of social science which seem relevant to the geographer's approach to spatial behaviour patterns.

The basic process of interaction in the schema is as follows. The *real world* is taken as the starting point, and it is represented as a source of *information*. The information content enters the individual through a system of *perceptual receptors*, and the precise meaning of the information is determined by an interaction between the individual's *value system* and their *image of the real world.* The meaning of the information is then incorporated into the image. On the basis of this information, the individual may require to adjust himself with respect to the real world. This requirement is expressed as a *decision* which can, of course, be one that involves no overt reaction. The links from the concept of a decision are two-fold (although these could be amalgamated). The first link is a re-cycling

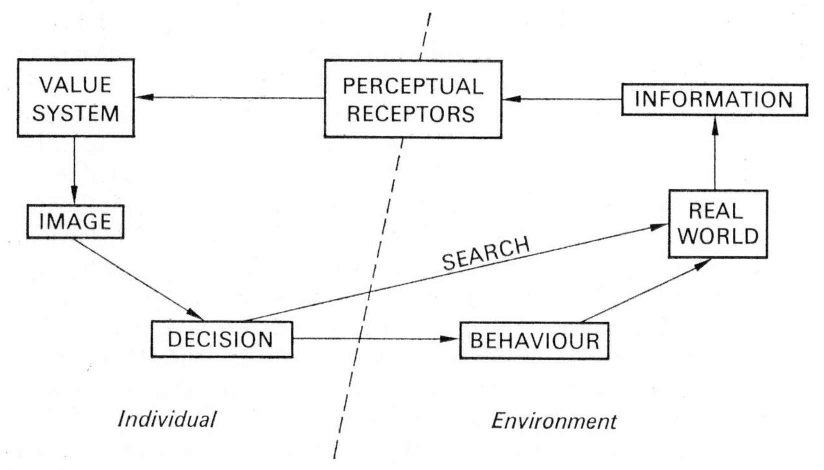

Fig. 8 A conceptual schema for research into geographic space perception.

process, called *search,* whereby the individual searches the real world for more information. This process can continue until either the individual decides that sufficient information has been acquired, or some time/cost limitation acts as a constraint to further search. A decision is then made which may be expressed as a pattern of *behaviour* which will in turn affect the *real world.* Since the real world undergoes a change, fresh information may result, and the whole process can continue. The schema therefore allows the space perception process to occur in a temporal as well as a spatial context.

We will now examine the fundamental theories and concepts which can be incorporated into the conceptual schema. The principal objective is *to view man as a decision maker,* an approach which is common to all of the social sciences. It is the wide range of highly developed decision theories that make this formulation attractive to the human geographer. Two particular applications of decision theory seem interesting. First, Davidson and others (1957, 3) argue that, 'even if a theory of rational decision has little general descriptive value, it may still have great interest as a normative theory.' Normative decisions theories can serve as a vital basis for comparing actual behaviour patterns with some form of expected pattern. In this sense they can be used to represent the constraints imposed by the real world, and therefore allow for a measure of the 'system performance' of human decisions to be obtained. This leads directly to the second issue that 'whether or not behaviouristics will provide the principal foundations for the new theories of decision making, it is clear that an understanding of behaviour is beginning to play a more and more important role in decision theories' (Shelley and Bryan, 1964, 25). Some of the developments in behavioural decision theory can be found in Edwards (1954; 1961), and it is in this context that we must view the concepts of 'satisficing' and 'optimising' developed by Simon (1957). These concepts have recently been introduced into geographic space perception by, amongst others, Wolpert (1964) and Kates (1962). The dual links from the decision box in the model allow for a related aspect of decision theory to be introduced. Search processes are intimately connected with the whole idea of perception, learning and behaviour. The theory of search behaviour may well be crucial in the development of theory linking the learning processes and perception. One example in the geographic literature is that of Golledge and Brown (1967) in their discussion of search processes and marketing behaviour. However, as Gould (1965) points out, as yet the relevance of search theory to human geography is limited. There has been much attention paid in psychology to the effect of limiting factors on the search and decision process, particularly the time/cost constraints (see Irwin and Smith, 1957, and Trull, 1966).

A second feature of the schema is the explicit inclusion of the idea that *behaviour is some function of the image of the real world.* In the past, this idea has remained somewhat of a platitude lurking in the background of the research process, from whence it could be produced for moral support as required. As Lucas (1964, 409) has said of a similar concept in the study of resources, 'all resources are defined by human perception. This has been said more often than used as an organizing concept in research.' Boulding (1956) uses the proposition that the image forms the basis for behaviour in a work which is notable as much for a demonstration of the far-reaching consequences of adopting this proposition as for the incisive discussions of the image itself. As yet, geographers are far more familiar with the behavioural manifestations of the formulation than with the image aspect, although Gould (1966, 1–4) cites examples of studies of spatial images. It is interesting to note that this idea was fundamental to the work of Koffka, a leading figure in the development of Gestalt psychology in the nineteen-thirties. Koffka (1935, 31) maintained that 'the relation between behaviour and the geographic environment must remain obscure without the mediation of the behavioural environment.' Kirk (1951, 1963) has developed Koffka's ideas in a spatial context and bases his formulation on the concept of man as a decision maker (but see the paper by Chein, 1954, for a discussion of the concept of a behavioural environment). However it is obvious that the image must remain the *focus* of research in geographic space perception, although we must refine the term to make it an operational concept.

The view of man embodied in the schema is one which is common to all social sciences. *Man is taken to be a complex information-processing system* (March and Simon, 1958) in which the information inputs are converted into messages which in turn act as the bases for decisions. The messages and the resultant images form the interface or point of contact between the individual and the environment: Beck (1967, 21) refers to this interface as ego space which 'is the individual's adaptation of observed to objective space, to produce a coherent and logically consistent view of sizes, shapes, and distances.' The idea basic to this view of the image/environment relationship is adaptation: that during the process of information intaking, there is a transformation process operating. Fortunately other social sciences, and psychology in particular, have studied this transformation process in detail. We will try to build some of their findings into our conceptual schema in order to clarify the relationships between the real world, the perceptual receptors, the value system and the image.

Bevan (1958, 34–7) in relating perception to behaviour theory, characterised the two principal views of perception as follows. In the first, perception is regarded as a *simple mediating process* in the transmission of information from the real world to the individual. Thus it represents a constant in the time/space dimension, and across people. This is basically a simple stimulus-response formulation, a mechanistic 'coin-in-the-slot' model that is familiar to geographers in the guise of Environmental Determinism. The assumption that a knowledge of the real world (the stimulus) is sufficient to predict the behaviour (the response) is demonstrably false. Bevan suggests that this view of perception has been replaced by that of a *complex interactive process*. In this formulation perception is a function of four sets of variables—the present stimulus input and the receptor functions (these form the basis of the simple mediating view), and two variables relating to motivation and past stimulation. We will call these last two sets of variables 'non-situational'. Katona (1951, 31) states that 'the basic scheme of psychological analysis is: situation—intervening variables—overt behaviour.' We will use the idea of intervening variables to characterise this view of perception. It is obvious that the image can be related to the variables which affect perception—in particular, the image would be heavily dependent on the past stimulation. Geographers are familiar with the intervening variable model as cognitive behaviourism (see the Sprouts, 1965, and Saarinen, 1966, 25–6). The *intervening variable formulation of perception* is the basis of our model, where information is transmitted through the perceptual receptors and the value system to form part of the image.

There is another view of perception that throws even more light on the process described in the preceding section. Forgus (1966, 10–12) presents a model of perception likening the organism or individual to a communication channel. There is an informational input into this communication channel in the form of stimuli, and the responses form the output. But there is rarely a one-to-one correspondence between the encoded input message (or information in the stimulus) and the decoded output message (or information in the perceptual response). The deviation from perfect correspondence is due to two sets of factors. The first is noise, which is stimulus energy blocking the information transmission, coming from either another input source or irrelevant sets in the organism. The second, and more important factor, is the existence of relevant sets in the organism which modify the encoded message before it is decoded. These are variable according to the characteristics of the organism, and therefore result in individual differences in perception. Obviously, these relevant sets are synonymous with the non-situational variables in our basic model.

The use of the communication channel model of perception has two advantages. First it permits us to use the concepts, theories and models of information theory. Thus, for example, Meier (1962) approaches the problem of urban growth dynamics via communications theory, and stresses the need for a constant intake of information to maintain urban growth. The general concepts of cybernetics have been extensively used in other cognate disciplines (see Deutsch, 1966, on political systems as control systems). However Medawar (1967, 37–58) suggests that the application of information concepts to organic systems may be a more difficult problem than people imagine, and a similar view has been expressed by Warr and Knapper (1968).

This communication channel model has a second advantage in that it lays stress upon these relevant sets or disturbing variables. Thus Bevan (1958, 35) argues that 'the identification and definition of the functions that influence [perception] become indispensable to an understanding and prediction of both perception and overt responses.' It is obvious that the concept called *value system* will have to be subjected to careful scrutiny. Already attempts are being made to identify the factors which cause people to view the same segment of the real world differently. We can view these factors as a set of filters. The perceptual receptors, forming the first battery of filters, are physiological, and not of interest to us as human geographers. From the point of view of spatial behaviour patterns, the set of psychological filters is more important in explaining both the patterns of behaviour and the variations that they display. These psychological filters include some form of Gestalt or pattern seeking function (see Kirk, 1963, and Beck, 1967, for discussions of the operation of this function). Similarly, the Whorfian hypothesis of linguistic relativity is alleged to operate, thus producing perceptual differences between language groups (but see Osgood, 1967, for contradictory evidence). Other filters include social class (Williamson, 1962), personal values (Postman *et al.*, 1948), value and need (Bruner and Goodman, 1947), and culture (Hall, 1959). These filters are not independent in their operation and will present complex problems in any analysis of their effects.

The final feature of the conceptual schema is that it is dynamic. We have already shown that the framework allows for a constant system of re-cycling of information, and therefore change in all elements. Our particular concern is with the change in the image as it affects spatial behaviour patterns. This permits the introduction of learning theory into the framework. At present, the only explicit use of learning theory in geography is that of Golledge (1967), who uses stochastic learning models to throw new light on central place theory and the temporal aspect of decision making.

We must now summarise the merits of the conceptual schema. Apart from making relationships explicit, it does identify the fundamental elements and indicate some of their potential interactions. Secondly, it is a dynamic model, focusing on the processes operating in space perception. But its most important feature lies in the direct attempt to relate human geography to the other social sciences.

3 The relation of attitude theory to the schema

We have dealt with the theory of perception almost exclusively, but attitude theory is also relevant. The parallels between the concepts of an image and an attitude are quite marked. For example, we assume that the knowledge of a person's image will allow us to predict his spatial behaviour: a similar claim has been made with respect to attitudes and behaviour. Yet as Fishbein (1967b, 477) has said:

> After more than seventy-five years of attitude research, there is still little, if any, consistent evidence supporting the hypothesis that knowledge of an individual's attitude toward some object will allow one to predict the way he will behave with respect to the object.

Fishbein's attempt to solve the problem posed by these findings is relevant to our conceptual framework in two ways. First, it throws more light on the nature of the concepts and their relationship. Second, it offers suggestions as to the value of past research into the perception and behaviour problem in geography, and points to the problems that we face in operationalising our conceptual framework.

Fishbein replaces the concept of an attitude by a formulation which can be diagrammed as in Figure 9. The fact that 'it is obvious that affect,

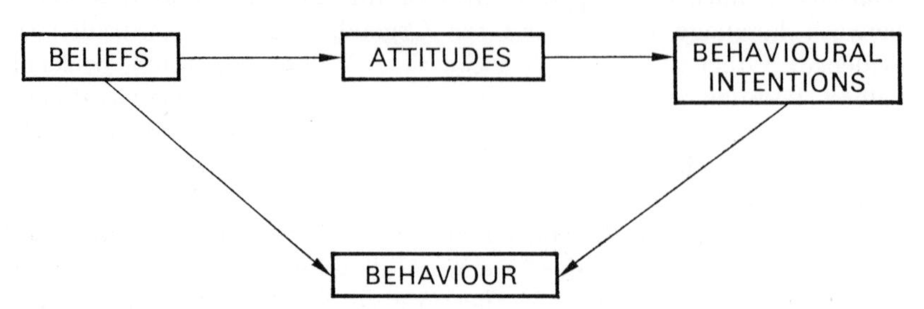

Fig. 9 Hypothesised components of an attitude.

cognition, and action are not always highly correlated' (Fishbein, 1967a, 257), leads to this more complex set of relationships. The relationship between attitude and behaviour is no longer a simple one, but is mediated by other factors.

This brief sketch barely does justice to an important series of papers. It is sufficient, however, to cause some concern to those interested in trying to produce predictive models for behaviour, based on perception and images. The concepts of belief and attitude are particularly relevant to our concept of an image. Fishbein (1967a) distinguishes between beliefs in the existence of an object, and beliefs about the nature of the object. Both sets of belief are expressed in terms of probability/improbability dimensions. Thus the research discussed in section II(1) under the heading of structure falls into line with the study of belief systems. Significantly, Boulding (1956, 5–6) refers to the image as being 'subjective knowledge' which 'largely governs my behavior.' This relationship may well, however, be both indirect and highly complex. Similarly, the work of the 'Chicago School' can be re-evaluated in the light of Fishbein's formulation. The questions that Kates (1967, 72–3) develops basically measure belief systems. On the basis of these responses, the intention is to infer reasons why people choose to locate in potentially hazardous areas. However Fishbein points out that attitudes, beliefs and behavioural intentions are often brought into line with behaviour. Consequently, Kates's approach seems to present great problems of causal relations and inferences. Once again, it demonstrates the need to be conversant with developments in other disciplines if research into geographic space perception is to make any real progress in theory development.

IV The problem of measurement in geographic space perception

Whiteman (1967, 28) distinguishes the 'scientific attitude', which is characterised by a readiness to analyse presuppositions and to renounce those that the whole state of affairs shows to be unwarranted, from the 'scientific method', which is based on the experimental testing of the consequences of hypotheses. Our approach to the problem of geographic space perception must comprehend both the scientific attitude and method, and these must dictate our general stance with regard to the conceptual schema developed in section III(2). However, to make any schema operational, we must introduce the 'scientific technique', which encompasses the particular means of testing hypotheses.

1 The problem of measurement

Kaplan (1964, 177) argues that 'measurement, in the most general terms, can be regarded as the assignment of numbers to objects (or events or situations) in accord with some rule.' Thus the property or attribute of the object we are measuring is represented in all future manipulations by the measure or number that we assign to it according to our pre-selected rule. Stevens (1959, 20–1) expresses the rationale behind the measurement procedure:

> Measurement is possible only because there is a kind of isomorphism between (1) the empirical relations among properties of objects and events and (2) the properties of the formal game in which numerals are the pawns and operators the moves.

The essential point is that we are dealing with pre-selected rules and a formal game. Viewed in this context, the measurement procedure is an arbitrary operation involving the nature of the objects, the purpose of the study, and the subjective evaluation of the observer. Measurement is a means to an end not an end in itself, and therefore measurements are made which allow certain hypotheses to be tested in certain ways.

There emerge three salient points about measurement. Although they apply in every scientific study, they are vitally important in attempting to grasp the problems involved in the measurement of geographic space perception. First, measurement procedures are defined by *fiat*: the observer develops a measurement technique to provide the data for testing particular hypotheses—i.e. 'purposive measurement', since the data required for research in geographic space perception is not usually immediately available. Nowadays, the range of measurement techniques is such that each individual researcher can use a standard technique and need not develop his own. Nevertheless, their arbitrary ad hoc nature remains. Second, as Stevens (1959, 29) argues, 'discouraging as it may appear, the outcome of "statisticizing" is no better than the empirical results that go into it.' Thus the whole procedure of both measurement and analysis is interrelated, and one cannot discuss one part without the other. Consequently we must concentrate on both data gathering (measurement) and data handling (analysis). Third, Kaplan (1964, 176) suggests that

> whether we can measure something depends, not on that thing, but on how we have conceptualised it, on our knowledge of it, above all on the skill and ingenuity which we can bring to bear on the process of measurement which our inquiry can put to use.

It is here that the crucial link between the conceptual framework and the measurement procedure is found. Torgerson (1958, 8) argues that the problem of *explication*—that is, the transformation of inexact conceptual constructs into exact ones which can be specified precisely and therefore measured—is vital in the development of theory:

> It is particularly acute in those disciplines that are in their initial stages of development. It is especially true at the present time in the social and behavioural sciences, where an immense amount of time has been devoted to the construction of complex and elaborate theoretical superstructures based on unexplicated, inexact constructs.

There have been several attempts to produce conceptual frameworks for the study of geographic space perception (see section 11(2)). However, until we can translate these into testable hypotheses, the development of theory is impossible.

One solution to the problem of explication is by the measurement procedure. 'Often, the problem of establishing a rule of correspondence for relating a construct to observable data reduces to the problem of devising rules for the *measurement* of the constructs' (Torgerson, 1958, 8). Thus Kaplan (1964, 177) argues that 'a procedure of measurement not only determines an amount, but also fixes what it is an amount *of*.' This solution has its problems. Webb and others (1966, 3) suggest that the mistaken belief in the operational definition of terms (that is, the confusion of the precision of carefully specified operations with operationalism by definitional fiat) 'has permitted social scientists a complacent and self-defeating dependence upon single classes of measurement—usually the interview or questionnaire.' This latter issue will be taken up in the following section. The fundamental problem of integrating conceptual frameworks and measurement procedures seems to be the most formidable obstacle in the development of theory in geographic space perception. If we do not make an extended effort to understand the nature of this relationship and its problems, we cannot hope to make any progress. It is not something to deter us however, for, as Kaplan states (1964, 25):

> Excessive effort can be diverted from substantive to methodological problems so that we are forever perfecting how to do something without ever getting around to doing it even imperfectly.

2 The problems of measuring images

The role of measurement, as discussed in the previous section, can be expressed as the crucial point at which ideas are translated into concrete data for later testing and analysis. Measurement thus serves to link the imagined or conceptual world with the real world. Having assigned this crucial role to measurement, we must look at some problems of measurement, particularly with respect to measuring images.

From the outset, we must recognise that we are dealing with a situation fundamentally different from that which a geographer conventionally has faced. Returning to our example of shopping trip behaviour, our previous attempts to explain such behaviour patterns have been based upon measurements drawn from the real world. Thus, in this case, we relate physical distances travelled in miles and the frequency of trip with the number of shops, their functional types (by some simple classification), and their locations, in miles, relative to the consumer. The concept of consumer space perception requires that we measure the evaluation of these shops and the transport media—that is, whether they are good or bad, clean or dirty, pleasant or unpleasant, etc. Such variables as pleasantness do not have any direct physical correlates, as, for example, does the type of shop variable. Therefore, in studying perception, we are trying to measure variables which exist primarily in the mind of an individual. Peterson (1967, 19) hypothesises a model of residential preference and argues that 'the model is not concerned with physical quantities that are objectively measurable. It is entirely contained within the subjective.' These variables are called '*psychological attributes*', and require measuring instruments which are fundamentally different from those normally employed by geographers.

However, we are fortunate that there is a wide range of techniques available, particularly in psychology. Craik (1968, 30) argues that

> [Psychology] has tended to concentrate its energies upon the study of basic, if apparently simple and inconsequential, processes and, perhaps, more importantly, upon the development of a repertory of quantitative methods and techniques appropriate to the phenomena it ultimately seeks to investigate and to understand.

He continues to list a wide range of techniques that are applicable to 'environmental psychology', a field which is synonymous with the interests of geographic space perception. Further evidence of the width of techniques available can be found in the books by Guilford (1954) and Nunnally (1967).

Although we are fortunate in being able to employ a wide range of measuring techniques, we must be aware of the many problems attendant to their use. We, as geographers, lack the accumulated wisdom of other social scientists familiar with the problems of the measurement of subjective data. Consequently our ability to assess the power of a measurement technique, its effectiveness, its range of applicability, and its reliability is very limited. To discuss the problems of measuring subjective data or psychological attributes, we will use examples drawn from attempts by geographers to measure spatial images.

There are two other major problems, that of *multiple operationism*, and that of *conceptualisation*. In section IV(2) we raised the point that the problem of explication via a measurement procedure can lead to an operational definition of a concept by only one type of measuring instrument. To a great extent, the instruments that have been developed to measure specific psychological attributes are based on such self-reporting devices as interviews or questionnaires, and these are susceptible to an extensive range of biases. At least nine different types of bias are listed by Webb and others (1966, 13–27). Therefore, they argue in favour of 'multiple operationism'. This procedure involves the use of a range of differing measuring instruments, some of which do not require observer/object interaction, but all of which are hypothesised to share in the theoretically relevant properties of the object, being measured, although having different patterns of irrelevant properties. Thus several measurement procedures are brought to bear on the same problem. Webb and others (1966, 4) suggest that this is necessary because we cannot specify the sources of variation which are operating in any situation with the same degree of precision as can physics or chemistry:

> In the social sciences, our measures lack such control. They tap multiple processes and sources of variance of which we are as yet unaware. At such a stage of development, the theoretical impurity and factorial complexity of every measure are not niceties for pedantic quibbling but are overwhelmingly and centrally relevant in all measurement applications which involve inference and generalisation.

The problem of biases is fundamental, particularly since we do wish to make inferences and predictions on the basis of the measurements. Although the range of 'non-reactive' measurement techniques is very limited, we should at least be aware of the problems involved in the use of 'reactive' measuring devices. Webb and others (1966, 173) stress that

> So long as one has only a single class of data collection, and that class is the

questionnaire or interview, one has inadequate knowledge of the rival hypotheses grouped under the term 'reactive measurement effects'.

We are concerned, therefore, with the sources of variation which can enter into a situation when we try to obtain measurements of psychological attributes. The problem of multiple operationism is that the measuring instrument and the object being measured (i.e. the individual) interact. The measures are therefore contaminated and may not be a reliable indicator of the underlying image that we wish to investigate. Thus the work of the 'Chicago School' is entirely dependent upon questionnaire/interview techniques. With the exception of Saarinen's (1966) study, these questionnaires have been of an ad hoc type. Similarly the work of Gould (1966) and White (1967) is open to criticism on the score of reactive measurement effects. In particular, the requirement that areas be rank-ordered according to 'residential desirability' leaves much scope for individual interpretation. The range of judgements could be attributable to the range of interpretations of the idea of residential desirability, when the object of interest was the variation in the ranking of the areas. Also the cognitive task of ordering 49 objects is beyond normal human ability, especially if one bears in mind that ordering requires a constant operation of the principle of transitivity. A similar set of biases occurs in the attempts to get people to draw their spatial images. Lee (1964) asked respondents to draw an enclosed line around a neighbourhood area on a map: this presupposes that mental space is areally continuous, and in addition, will be dependent upon individual graphic ability and familiarity with the map format as a means of representing space. These other sources of variation may account for the discrepancy between his maps and those of Lynch (1960).

The problems of this approach to measurement are further complicated by the findings of Fendrich (1967). He demonstrated that 'the definition of the measurement situation influences the way respondents express their attitudes' and that therefore 'verbal attitudes can be either consistent or inconsistent with overt behaviour, depending upon the way respondents define the attitude measurement situation' (Fendrich, 1967, 355). Thus we must be constantly aware of both the sources of variation that can affect our measurement situation, and the inferences drawn from the data. There is no immediate palliative, except constant vigilance and healthy scepticism. Fortunately this is a problem common to all of the social sciences, and they have not found it an insuperable difficulty.

The second problem in measuring psychological attributes concerns 'conceptualisation'. This can best be explained by the following quotation from a recent article by Burrill (1968, 4):

the Atlantic coast swamp with all those attributes is a sort of complex, the nature of which is clearly known to the local people who are parties to a tacit understanding about the usage of the term. We who were not parties to the basic regional understanding had been trying to categorise this kind of entity in terms of a single kind of attribute rather than in terms of a multi-attribute feature for which we had no pigeon hole, no term, no concept.

Burrill was concerned with problems of mapping and the definition of mappable entities. But this is a problem of conceptualisation. Groups of people have different images of the real world. This is one of the areas of concern and a problem of geographic space perception measurement. How can we be sure that the entity we are attempting to measure has an existence, and hence meaning, to those people from whom we are collecting data? In Burrill's example the same object was viewed differently by two groups, the local inhabitants and the map-makers. Consequently there was a problem of communication, since their conceptualisations of the real world differed radically.

This has a parallel in the problem of 'value transference' in historical geography (see Kirk, 1952; 1963). The Narrolls (1963) discuss the problem in an anthropological, cross-cultural context. But it also applies within cultures, as we have shown above. Different groups, defined according to different criteria, will have different viewpoints. (This is an underlying assumption of current research into geographic space perception.) Therefore a member from one group will have difficulty in measuring the images of another group. The real problem is that we, as geographers, represent a group with a particular viewpoint or 'spatial style' (see Beck, 1967). Our way of approaching the world is based on the concepts of relative location, proximity and distance, and especially geared to the use of maps. Boulding (1956, 65) has argued that 'the map itself . . . has a profound effect on our spatial images.' He may be correct. It is also possible that our way of thinking may not be in accord with people in general. The question of the nature of images of the real world is therefore crucial, and will require some extremely sensitive measuring devices. Above all, in our research into perception, we must beware of imposing upon people ways of thought and conceptualisation foreign to them.

Despite these fundamental measurement problems, the recent literature on perception in geography demonstrates an increasing awareness of the power and applicability of existing psychological measurement techniques. Saarinen (1966) used the thematic apperception test (TAT) to depict the images which farmers associated with the 'dustbowl' conditions of the American Plains in the nineteen-thirties. The remarkable success of this technique offers a promising lead for future research. Both Peterson (1967) and Sonnenfield (1967) used visual presentations of material to elicit responses, an approach which also forms the basis of the TAT. Peterson used a variant of the 'paired comparisons technique' with respect to photographs of different residential environments in order to estimate the underlying dimensions of residential preference. In addition to a photo-slide scaling test, Sonnenfeld employed a rating scale, the semantic differential. This scaling technique offers great potential for obtaining quantitative, and therefore comparable, reconstructions of images of objects.

It is interesting to note that these measurement techniques have been used principally to arrive at measurements of images. In the conceptual schema the image was distinguished from the real world, and so we must also provide methods for the measurement of the real world. In his discussion of the three basic kinds of spaces, Beck (1967, 21) refers to the real world as objective space—'the space of physics and mathematics, measured by universal standards along dimensions of distance, size, shape, and volume.' We must therefore agree upon measuring techniques for the real world in order that our comparisons between the image and the real world are meaningful and consistent. One range of models which may be useful is that of normative economics, especially the models developed in operations research. The value of such models was shown by Wolpert (1964) who used a linear programming model to describe the potential universe of action open to Swedish farmers. Against this he could match their actual decisions.

It is obvious that the measurement problem is one which will trouble human geographers interested in perception. The prerequisite would seem to be an extremely close liaison and contact with those people having the greatest degree of familiarity with this problem, namely the psychologists. It is significant, therefore, that two of the most successful studies of geographic space perception, those of Sonnenfield (1966; 1967) and Saarinen (1966), have been based on a close contact with psychologists.

3 Measurement procedures: the use of multidimensional scaling models

Our earlier distinction between data gathering (measurement) and data handling (analysis) forms the basis for the discussion of the measurement procedure. It is a distinction made by Selltiz and others (1965, 147) who categorise the measurement procedure as consisting of:

1 a technique for collecting data
2 a set of rules for using the data.

It is logical to consider these two requirements as stages in an interrelated process whereby the analytic model (which determines the set of rules for using the data) requires a certain type of data, whilst the data determines the type of analytic model which is applicable. Consequently, although we will discuss the measurement procedures in two stages, in any practical application the two stages cannot be approached independently.

The relationships within the measurement procedure are expressed in Figure 10, which is partially based on Coombs (1964, 4). The objective of the measurement procedure is the development of theory through the

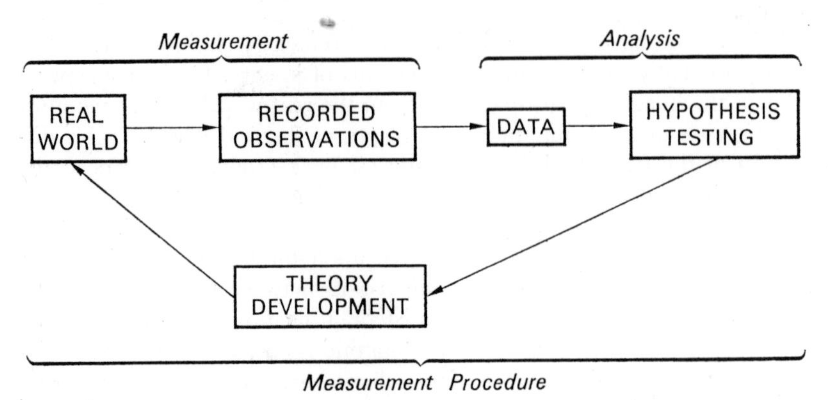

Fig. 10 The measurement procedure.

testing of hypotheses. The developed theory then serves as a picture of the real world. Medawar (1967, 89) sums the process up thus:

> Devising a hypothesis is a 'creative art' in the sense that it is the invention of a possible world, or a possible fragment of the world; experiments are then done to find out whether or not the imagined world is, to a good enough approximation, the real one.

The class of measurement techniques that seems most appropriate are multidimensional scaling models where 'pyschological scaling methods are procedures for constructing scales for the measurement of psychological attributes' (Torgerson, 1958, ix). It is significant that scaling models have been developed such that both the measurement and analysis stages are incorporated into one generalised model. Thus:

> Scaling models have come to be associated with the particular logical systems that lead to geometrical representations of inferential classification—for example, unidimensional scales and multidimensional spaces. (Coombs, 1964, vii–viii)

Measurement was earlier defined as the procedure of attaching numbers to the properties of the objects so that the relations between the numbers matched the relations between the properties of the objects. In multidimensional scaling models, the assigned numbers are used to locate or to map the objects as points in a space. The analysis or later manipulation is performed upon these points in the multidimensional space.

Guilford (1954, 246) suggests that multidimensional scaling models should be employed in the following situations:

1 with complicated stimuli (where stimuli are the objects of concern)
2 with stimuli whose physical dimensions are not well known
3 with judgements of psychological qualities for which there are no recognised corresponding physical dimensions.

The basic concepts of geographic space perception fall within all three categories: for example, an image is complex and multifaceted, its relation to the physical real world uncertain, and it has no direct physical expression.

The two fundamental considerations in the use of multidimensional scaling models are the nature of the measurements, and the nature of the space, where the space is a formal geometrical model which constrains the analysis. The measurements are obtained by using a technique which extracts relevant information from the real world.

However, the principal feature of multidimensional scaling models lies in the process of mapping the measures into a space. The space forms a model for the relations between objects. Conventionally two basic characteristics of the space are of interest to us: these are its structural dimensionality, and the relative locations of the objects in the space. Thus Torgerson (1958, 247–8) states the typical problem which could be handled by a multidimensional scaling model: given a set of objects, varying with respect to an unknown number of dimensions, determine:

1 the minimum dimensionality of the set
2 the projections of the objects (the scale values) on each dimension.

This underlying approach is already familiar to geographers through the recent use of the factor analysis model, where the values of a set of objects on a series of scales form the basic measurement. The space of the original set of scales is then reduced to a smaller dimensionality, and the scale values or factor scores of the objects on the new dimensional structure are computed.

However the space of the factor analysis model is only one of many which are available. The value of this range of spaces will increase as the empirical research into geographic space perception begins to yield results. These results will assist in the choice of the best spatial model for the nature of the particular problem. Coombs (1964, 16) states that:

> The space is assumed to be a metric space, but the distance function is not specified. Most psychological scaling models assume a multidimensional psychological space to be Euclidean, to incorporate the familiar everyday notion of distance. Some do not even require a metric space, whereas others specify a distance function other than Euclidean . . .

We will briefly discuss two of the principal flexibilities of the various spatial models—first the idea of metric as against non-metric spaces, and second, the geometrical model. (For a more detailed discussion see Bassett and Downs, 1968.)

The so-called 'non-metric' approach was developed by Shepard (1962a and b) and has been progressively extended by, among others, Kruskal (1964a and b) and Shepard (1966). The distinction between the non-metric and the metric approaches lies in the nature of the measurement scale underlying the measurements. A non-metric approach will operate on all types of ordinal scale (that is, a simple rank order scale and a higher ordered metric scale, which contains measures of all $n(n - 1)$ relations between n points). The metric approaches require data at an interval scale level of measurement. It has been demonstrated, however, that the results obtained by performing a non-metric data analysis approximate so closely to those that would be obtained by using a metric analysis that the two are virtually interchangeable. Not only does this avoid the problem of arbitrarily producing an interval scale for measuring these psychological attributes, but it also avoids constraints on the nature of the function used for mapping the measures as points in the space. The mapping function for a non-metric space is a monotonic one, whereas the metric models (such as factor analysis) assume a linear mapping function. However, the exact relations between these two types of analysis are highly complex and imperfectly understood, particularly since the development of non-metric approaches is still continuing.

The nature of the underlying geometry is also relevant in the choice of a model. Essentially we are trying to match the relationships between the objects with their representation as distances in a space. Consequently the underlying geometry is vital in that it affects the calculation of the distances between the points. Reichenbach (1958, 6) states:

> After the discoveries of non-Euclidean geometries the duality of *physical* and *possible* space was recognised. Mathematics reveals the possible spaces; physics decides which among them corresponds to physical space.

Therefore, we as geographers must decide which possible space fits the concepts of geographic space perception. The importance of this choice is particularly relevant in perception studies since psychologists have argued that many aspects of perception require non-Euclidean geometries to represent them adequately (see Luneberg, 1947; Corcoran, 1966). Attneave (1950) proposed the adoption of the 'city-block' metric with respect to visual perception. This example will serve to demonstrate that the choice of the geometry affects the calculation of the distance between points in the various possible spaces. In Euclidean geometry, the distance between two points, j and k, is defined as:

$$d(j,k) = \left[\sum_{m=1}^{t} (| a_{jm} - a_{km} |)^2 \right]^{\frac{1}{2}} \qquad (j,k) = 1, 2, \ldots n$$

where j,k are alternative subscripts for points j,k; $d(j,k)$ is the distance between points j and k; m is the subscript for orthogonal axes of the space ($m = 1, 2, \ldots, t$); and a_{jm} is the projection of stimulus j on axis m. Kruskal (1964a, 22–4) generalises this equation in terms of distance functions known as *Lp*-norms of Minkowski *r*-metrics:

$$d(j,k) = \left[\sum_{m=1}^{t} (| a_{jm} - a_{km} |)^r \right]^{1/r}$$

Thus the *r*th power and the *r*th root replace squaring and the square root. In the case of the city-block metric, *r* becomes 1. The distance between two points *j* and *k* in the two spaces is calculated as in Figure 11.

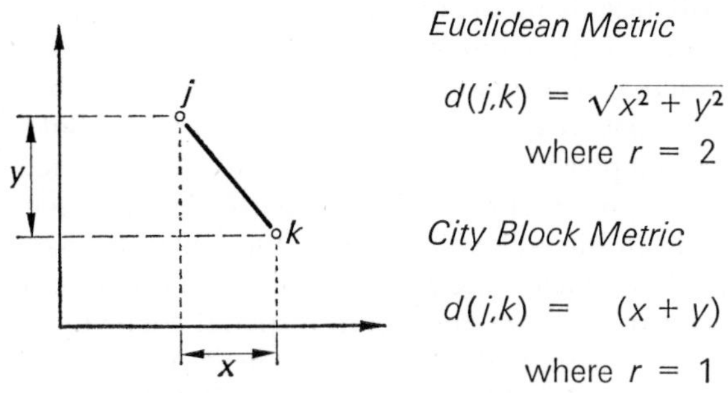

Fig. 11 The Euclidean and the city-block metrics.

The measurement procedure is therefore full of decision points for the researcher. First, the particular field of concern has to be isolated. Then, after establishing hypotheses, the relevant variables must be explicated, usually via the choice of a particular set of measurement techniques. If possible these must be selected with the concept of 'multiple operationism' in mind. In the analysis stage the type of model must be selected, while careful attention is paid to the assumptions that it makes and hence imposes on the measurements. Consequently the development of theory in geographic space perception will be dependent upon the skill and knowledge of the researcher in carrying out the measurement procedure. Yet we must be aware of the decisions, implicit and explicit, if we are not to make grievous mistakes.

V Conclusion

The objective of the behavioural revolution in modern geography is to provide new and more powerful explanations of spatial behaviour patterns by employing more realistic conceptions of man's physiological and psychological limitations. Within this approach lies the concept of geographic space perception which is concerned with the image, the mental or cognitive space within which an individual operates. However, this formulation is not in itself sufficient to ensure that the study of geographic space perception will occupy a key role in the current attempt to develop geographic theory.

Of the areas which will be crucial in the development of research into perception, we have stressed first the need for an overall conceptual approach which will permit us to integrate with other social sciences, and second the fundamental measurement problem. The need for a close liaison and integration with cognate social science disciplines has been recognised in the last few years in our work on perception. But it is a trend that we must emphasise and extend so that we have adequate channels of communication with psychology in particular. In essence the study of perception is an inter-disciplinary one. Our role as geographers will be the familiar one of integrating and synthesising findings from other disciplines but always using these findings in a spatial context.

The second area of concern is that of measurement. We face substantially different measurement problems to those previously encountered in geography. This may well prove to be a major stumbling block because no matter how sophisticated the theoretical superstructures that we erect, the acid test remains in their testing—and the test depends upon our ability to measure successfully. Nunnally (1967, 5) goes as far as to suggest that 'major advances in psychology, and probably in all sciences, are preceded by breakthroughs in measurement methods.' Consequently, we must pay particular attention to the problem of the measurement of psychological attributes. Once again this involves drawing upon the expertise of other disciplines.

Despite these obvious problems, perception remains an intellectually stimulating and potentially rewarding branch of human geography. But we must not minimise or avoid the obstacles, particularly those of measurement, if research into perception is to become an important area and not just a popular one.

VI Acknowledgements

I would like to acknowledge the contributions of the many people who, by their comments, both appreciative and condemnatory, have assisted in moulding the final form of this paper. These include my colleagues at Bristol, Keith Bassett, Andrew Cliff, Robert Colenutt, Stewart Cowie, Professor Peter Haggett, Dr David Harvey and Rodney White, and those who attended seminars given at the Departments of Geography in Exeter University and University College, London. These seminars were valuable proving grounds for some of the ideas in this paper. As always, however, the final sins of omission and commission remain my sole responsibility.

VII References

Appleyard, D., Lynch, K. and **Myer, J.** 1964: *The view from the road.* Cambridge, Massachusetts: MIT Press.

Attneave, F. 1950: Dimensions of similarity. *American Journal of Psychology* 63, 516–56.

Bassett, K. and **Downs, R. M.** 1968: *The use of multidimensional spatial models in geographic research.* University of Bristol, Department of Geography: unpublished manuscript.

Beck, R. 1967: Spatial meaning and the properties of the environment. In Lowenthal, D., editor, Environmental perception and behavior, *University of Chicago, Department of Geography, Research Paper* 109, 18–41.

Berry, B. J. L. and **Marble, D. F.** 1968: *Spatial analysis.* Englewood Cliffs, N.J., and London: Prentice Hall. (512 pp.)

Bevan, W. 1958: Perception: evolution of a concept. *Psychological Review* 65, 34–55.

Boulding, K. E. 1956: *The image.* University of Michigan Press. (184 pp.)

Bruner, J. S. and **Goodman, C. C.** 1947: Value and need as organizing factors in perception. *Journal of Abnormal and Social Psychology* 42, 33–44.

Burrill, M. F. 1968: The language of geography. *Annals of the Association of American Geographers* 58, 1–11.

Burton, I. 1963: The quantitative revolution and theoretical geography. *Canadian Geographer* 7, 151–62.

Burton, I. and **Kates, R. W.** 1964: The perception of natural hazards in resource management. *Natural Resources Journal* 3, 412–41.

Chein, I. 1954: The environment as a determinant of behaviour. *The Journal of Social Psychology* 39, 115–27.

Coombs, C. H. 1964: *A theory of data.* New York and London: Wiley.

Corcoran, D. W. J. 1966: A test of some assumptions about psychological space. *American Journal of Psychology* 79, 531–41.

Craik, K. H. 1968: The comprehension of the everyday physical environment. *Journal of the American Institute of Planners* 34, 29–37.

Davidson, D., Suppes, P. and **Siegel, S.** 1957: *Decision making; an experimental approach.* Stanford University Press.

Deutsch, K. W. 1966: *The nerves of government; models of political communication and control.* New York: Free Press of Glencoe; London: Collier-Macmillan. (304 pp.)

Edwards, W. 1954: The theory of decision making. *Psychological Bulletin* 51, 380–417.

1961: Behavioral decision theory. *Annual Review of Psychology* 12, 473–498.

Fendrich, J. M. 1967: A study of the association among verbal attitudes, commitment and overt behavior in different experimental situations. *Social Forces* 45, 347–55.

Fishbein, M. 1967a: A consideration of beliefs, and their role in attitude measurement. In Fishbein, M., editor, *Readings in attitude theory and measurement,* New York and London: Wiley, 257–66. (512 pp.)

1967b: Attitude and the prediction of behavior. In Fishbein, M., editor, *Readings in attitude theory and measurement,* New York and London: Wiley, 477–92. (512 pp.)

Forgus, R. H. 1966: *Perception.* New York and London: McGraw-Hill. (402 pp.)

Golledge, R. G. 1967: Conceptualizing the market decision process. *Journal of Regional Science* 7(2) (Supplement), 239–58.

Golledge, R. G. and **Brown, L. A.** 1967: Search, learning, and the market decision process. *Geografiska Annaler* 49*B*, 116–24.

Gould, P. R. 1965: *A bibliography of space searching procedures for geographers. Pennsylvania State University, Department of Geography, Research Note.*

1966: On mental maps. *Michigan Inter-University Community of Mathematical Geographers, Discussion Paper* 9.

1967: Structuring information on spacio-temporal preferences. *Journal of Regional Science* 7(2) (Supplement), 259–74.

Guilford, J. P. 1954: *Psychometric methods.* New York and London: McGraw-Hill.

Haggett, P. 1965: *Locational analysis in human geography.* London: Arnold, New York: St Martin's Press. (352 pp.)

Hall, E. T. 1959: *The silent language.* New York: Doubleday.

Harvey, D. W. 1966: Geographical processes and the analysis of point patterns. *Transactions of the Institute of British Geographers* 40, 81–95.
Heinemeyer, W. H. 1967: The urban core as a centre of attraction. In University of Amsterdam, Sociographical Department, *Urban core and inner city*, Leiden: Brill, 82–99.
Howard, I. P. and **Templeton, W. B.** 1966: *Human spatial orientation.* New York and London: Wiley. (542 pp.)
Huff, D. L. 1959: Geographical aspects of consumer behaviour. *University of Washington Business Review*, June, 27–38.
1960: A topographical model of consumer space preferences. *Papers and Proceedings of the Regional Science Association* 6, 159–173.
Irwin, F. W. and **Smith, W. A. S.** 1957: Value, cost, and information as determiners of decision. *Journal of Experimental Psychology* 54, 229–32
Kaplan, A. 1964: *The conduct of inquiry.* San Francisco: Chandler.
Kates, R. W. 1962: Hazard and choice perception in flood management. *University of Chicago, Department of Geography, Research Paper* 78.
1967: The perception of storm hazard on the shores of Megalopolis. In Lowenthal, D. editor, Environmental perception and behavior. *University of Chicago, Department of Geography, Research Paper* 109, 60–74.
Katona, G. 1951: *Psychological analysis of economic behavior.* New York and London: McGraw-Hill.
Kirk, W. 1951: Historical geography and the concept of behavioural environment. In *Silver Jubilee Souvenir and N. Subrahmanyam memorial volume.* Madras: Indian Geographical Society, 152–60.
1963: Problems of geography. *Geography* 47, 357–71.
Koffka, K. 1935: *Principles of gestalt psychology.* London: Kegan Paul. (704 pp.)
Kruskal, J. B. 1964a: Multidimensional scaling by optimizing goodness-of-fit to a nonmetric hypothesis. *Psychometrika* 29, 1–28.
1964b: Nonmetric multidimensional scaling: a numerical method, *Psychometrika* 29, 115–29.
Lee, T. R. 1964a: Urban neighbourhood, 1: Socio-spatial schemata and contemporary planning. University of St Andrews: Department of Psychology (Mimeo.)
1964b: Psychology and living space. *Transactions of the Bartlett Society* 2. 9–36.
Lewin, K. 1963: *Field theory in social science.* London: Tavistock Publications (366 pp.)
Lowenthal, D. 1961: Geography, experience, and imagination: towards a geographical epistomology. *Annals of the Association of American Geographers* 51, 241–60.

Lucas, R. C. 1964: Wilderness perception and use: the example of the Boundary Waters canoe area. *Natural Resources Journal* 3, 394–411.

Luneburg, R. K. 1947: *The mathematical analysis of binocular vision.* Princeton University Press.

Lynch, K. 1960: *The image of the city.* Cambridge, Massachusetts MIT Press and Harvard University Press.

Marble, D. F. 1959: Transport inputs at urban residential sites. *Papers and Proceedings of the Regional Science Association* 5, 253–66.

March, J. G. and **Simon, H. A.** 1958: *Organizations.* New York and London: Wiley. (262 pp.)

Medawar, P. 1967: *The art of the soluble.* London: Methuen. (168 pp.)

Meier, R. L. 1962: *A communications theory of urban growth.* Cambridge, Massachusetts: MIT Press.

Naroll, R. and **Naroll, F.** 1963: On bias of exotic data. *Man* 25, 24–6.

Nunnally, J. C. 1967: *Psychometric Theory.* New York and London: McGraw-Hill. (640 pp.)

Olsson, G. 1967: Geography 1984. *University of Bristol, Department of Geography, Seminar Paper Series A* 7.

Osgood, C. E. 1967: Cross-cultural comparability in attitude measurement via multilingual semantic differentials. In Fishbein, M., editor, *Readings in Attitude Theory and Measurement*, New York and London: Wiley, 108–16. (512 pp.)

Peterson, G. L. 1967: A model of preference: quantitative analysis of the perception of the visual appearance of residential neighbourhoods. *Journal of Regional Science* 7, 19–31.

Postman, L., Bruner, J. S. and **McGinnies, E.** 1948: Personal values as selective factors in perception. *Journal of Abnormal and Social Psychology* 83, 148–53.

Reichenbach, H. 1958: *The philosophy of space and time.* New York: Dover; London: Constable. (295 pp.)

Research and Design Institute 1967: *Directory of behaviour and environmental design.* Providence, Rhode Island.

Rostlund, E. 1956: Twentieth century magic. *Landscape* 5, 23–6.

Saarinen, T. F. 1966: Perception of the drought hazard on the Great plains. *University of Chicago, Department of Geography, Research Paper* 106.

Selltiz, C., Jahoda, M., Deutsch, M. and **Cook, S. W.** 1965: *Research methods in social relations.* London: Methuen. (638 pp.)

Shelley, M. W. and **Bryan, G. L.,** editors, 1964: *Human judgements and optimality.* New York: Wiley.

Shepard, R. N. 1962a: The analysis of proximities: multidimensional scaling with an unknown distance function, I. *Psychometrika* 27, 125–39.

1962b: The analysis of proximities: multidimensional scaling with an unknown distance function, II. *Psychometrika* 27, 219–46.

1966: Metric structures in ordinal data. *Journal of Mathematical Psychology* 3, 287–315.

Simon, H. A. 1957: *Models of man: social and rational.* New York and London: Wiley. (287 pp.)

Sonnenfeld, J. 1966: Variable values in space and landscape: an inquiry into the nature of environmental necessity. *Journal of Social Issues* 22, 71–82.

1967: Environmental perception and adaptation in the Arctic. In Lowenthal, D., editor, Environmental perception and behavior. *University of Chicago, Department of Geography, Research Paper* 109, 42–59.

Sprout, H. and **Sprout, M.** 1965: *The ecological perspective on human affairs with special reference to international politics.* Princeton University Press and Oxford University Press. (248 pp.)

Steinitz, C. 1967: *Congruence and meaning: the influence of consistency between urban form and activity upon environmental knowledge.* Harvard University: Department of City and Regional Planning. (Mimeo)

Stevens, S. S. 1959: Measurements, psychophysics, and utility. In Churchman, C. W. and Ratoosh, P., editors, *Measurement: definitions and theories*, New York and London: Wiley. (274 pp.)

Torgerson, W. S. 1958: *Theory and methods of scaling.* New York and London: Wiley. (460 pp.)

Trowbridge, C. C. 1913: On fundamental methods of orientation and imaginary maps. *Science* 38, 888–97.

Trull, S. G. 1966: Some factors involved in determining total decision success. *Management Science* 12, B-270–B-280.

Warr, P. B. and **Knapper, C.** 1968: *The perception of people and events.* New York and London: Wiley.

Webb, E. J., Campbell, D., Schwartz, R. and **Sechrest, L.** 1966: *Unobtrusive measures: non-reactive research in the social sciences.* Chicago: Rand McNally; London: Eurospan. (240 pp.)

Webber, M. 1964: Culture, territoriality, and the elastic mile. *Papers and Proceedings of the Regional Science Association* 14, 59–70.

White, G. F. 1945: Human adjustment to floods. *University of Chicago, Department of Geography, Research Paper* 29.

White, R. R. 1967: *Space preference and migration: a multidimensional analysis of the spatial preferences of British school-leavers, 1966.* Pennsylvania State University: unpublished Master's thesis.

Whiteman, M. 1967: *Philosophy of space and time.* London: Allen and Unwin. (436 pp.)

Williamson, R. C. 1962: Social class determinants of perception and adjustment in an adolescent and adult sample: El Salvador. *Journal of Social Psychology* 57, 11–21.

Wolpert, J. 1964: The decision process in spatial context. *Annals of the Association of American Geographers* 54, 537–58.

1965: Behavioral aspects of the decision to migrate. *Papers and Proceedings of the Regional Science Association* 15, 159–69.

Building models of urban growth and spatial structure

by Robert J. Colenutt

Contents

I Introduction[1]

1 The problem

FEW problems appear more disturbing and pressing to both Western societies and the developing nations, than the pace of urbanisation and the growth of cities and city systems. As the crisis looms, it is apparent that relatively little information and few insights are available that might suggest how urban growth can be regulated, organised, or indeed, how new urban settlements might be planned. The social sciences, and particularly geography, have barely begun to tackle the myriad of intellectual and social problems that the cities are generating. Geographers have only belatedly begun to show interest or assume responsibility for examining urban growth and urban spatial structure. This function has been usurped quite naturally by city and regional scientists from other disciplines who have accumulated a variety of tools, techniques, concepts and models that are extremely relevant to geographical analysis. In this paper, we shall look at some of these approaches and models and attempt to relate them to the problem of the explanation and understanding of the metropolis.

First, the problem of urban growth must be more closely defined. In this paper we are focusing only on the *locational* aspect of the problem which could be more broadly described as the changing spatial pattern of land uses and activities in the city. We are thus interested in investigating *how* land use patterns have evolved and *what form* future changes are likely to take. In effect the two end-products of our analysis are (1) a set of concepts, ideas and models that provide insights into the behaviour of the urban system, and from these, (2) a set of models that will permit experimentation with urban form and structure and prediction of the future spatial pattern and operation of the city.

There appear to be two ways of tackling the problem put forward above. The first is to derive axioms and theories about the processes and patterns of locational change in the city. Using these fundamental conceptual insights, *deductive models*, defined by Harvey (1966, 552) as the formal presentation of theories, can be constructed. But these models may not be capable (because of the restricted nature of the theory or the difficulties of empirical testing) of producing the other end-product, experimental models. If prediction and experimentation are important, and in urban planning they are critical, it is essential that techniques are available

[1] During the time between the writing and printing of this article, a considerable amount of work has been produced in this field. Anyone requiring details of new literature arising from this may apply to the author, at the Department of Geography, Syracuse University.

even though they may not be models in a deductive sense. It is not a necessary nor a sufficient condition that these models (or any other type of model for that matter) are *explanatory*, only that they are accurate *descriptive and analytical* tools. Hence, both types of model will be examined in this paper and considerable attention paid to the interdependencies and usefulness of the two approaches.

2 The scope of the paper

Four aspects of the model building process will be used to illustrate the differences and usefulness of the two types of model discussed above.

1 The theoretical content of urban growth models
2 The problem of aggregation
3 Dealing with time
4 Choosing a mathematical structure for the model.

These aspects represent problems that the model builder encounters during the model design process. In some ways, they also act as constraints on the design and are instrumental in determining whether the model is theoretically elegant or predictively precise (or both).

Most of the illustrations and examples will be drawn from work on residential and retail location models, since these have received far more attention than models for other sectors such as intra-urban industrial location (see Putnam, 1967; Steger, 1964) and office location (see Cowan and Ireland, 1967).

II The theoretical content of urban growth models

In this section, we shall look first of all at the body of 'theories' and concepts that are available to the model builder, and discuss the extent to which these ideas have either been adopted in urban models or proved useful (see Table 1). Secondly, some suggestions will be made about the reasons for the low theoretical content of many urban models and the difficulties of making them more realistic simulators of urban growth and development.

1 Concepts from urban ecology

The best known 'theories' in urban ecology are essentially descriptive. The Burgess (1927) model of the concentric ring form of urban spatial structure interprets city growth as the product of centrifugal social forces

Table 1 The use in urban models of conceptual developments in urban research

Discipline	Concept	Principal author(s)	Derivative mathematical model	Examples of application in urban models
Urban ecology	Concentric zone	Burgess	None } Descriptive only	None
	Sector theory	Hoyt	None } Descriptive only	None
	Multiple nuclei	Ullman	None } Descriptive only	None
Sociology and social physics	Gravity and potential concepts	Zipf, Stewart	Gravity and potential models	Hansen's accessibility model Lakschmanan–Hansen retail model Huff retail model
	Intervening opportunity	Stouffer, Schneider	Intervening opportunity model	Harris retail model Hamburg Niagara frontier model
Geography	Central place theory	Lösch, Berry	No general model	Berry retail model
	Behavioural concepts—searching, learning, etc.	Golledge, Brown	Mathematical learning models (Markov models)	No operational model
Urban economics	Theories of the operation of the urban land market	Haig, Ratcliffe, Wendt	No general model	Alonso model Wingo model Herbert-Stevens model

but does not make explicit the precise reasons for this evolution. It points to some of the processes but does not attempt to isolate the variables or postulate a mathematical model.

Similarly, Hoyt's concept of the sector growth of urban development (1939) and Harris and Ullman's (1945) multiple nuclei description of the city do not specify the variables or factors, decision makers, or causal relationships that might be responsible for city growth and form. Moreover the concepts allude primarily to sociological forces and do not take into account economic conditions such as the behaviour of the urban land market, although Firey (1960) and Hoyt do recognise these conditions indirectly. Finally the theories, if they can be called that, have not suggested or been responsible for the construction of any operational analytical model. Leven's recent (1967) discussion of the concentric ring model, for example, is basically descriptive and is not phrased in mathematical model terms. The theories thus remain descriptive and graphic, and contrast with those concepts put forward by sociologists which have been widely adopted in urban models.

2 Concepts from sociology and social physics

The major contribution of the social physicists was the adoption of the gravity concept from Newtonian physics in models of social interaction. The interactance model developed by Zipf (1947), Stewart (1950) and others, and discussed by Carrothers (1956), Schneider (1959), and more recently by Catton (1965) has proved to be not only a useful descriptive device but a predictive tool of some power.

The model was taken from physics and has undergone little fundamental amendment since Zipf, and despite criticisms of the model on theoretical grounds by Huff (1965) and others, the model has been applied widely in area transportation studies both for predicting changes in land use and for traffic distribution over a transport network (United States Department of Commerce, 1962; 1963). For land use forecasts, these studies have often used Hansen's version of the gravity model, known more generally as the accessibility model (Hansen, 1959, 75). The accessibility model expresses residential growth as a function of the availability of vacant land and proximity to employment. The general form of the model is as follows:

$$G_i = G_t \frac{A_i^a \, V_i}{\sum_{t=1}^{n} A_i^a \, V_i} \qquad (1)$$

where G_i is the growth forecast for zone i; G_t is the total regional growth;

A_i is the accessibility index for zone i; V_i is the vacant land in zone i, and a is an empirically derived exponent; and the accessibility index, A_i, is written as:

$$A_i = \sum \frac{E_j}{T_{ij}^b} \qquad (2)$$

where E_j is the measure of activity (for example, total employment) in zone j; T_{ij} is the travel time (generally the shortest path) from zone i to zone j; b is an empirically derived exponent.

Several modifications have been introduced into the model but they have not substantially altered the predictive power or theoretical elegance of the basic gravity formulation. Stouffer (1940) introduced intervening opportunities into the model and this was operationalised by Schneider (1959) as a trip distribution model. Lathrop and Hamburg (1965) have taken Schneider's opportunity model and used it for modelling residential development. This opportunity model views the spatial distribution of an activity as the successive evaluation of alternative opportunities for sites which are rank-ordered in time from an urban centre where the opportunities are defined as the product of available land and the density of activity (Lathrop and Hamburg, 1965, 96). Apart from the intervening opportunity addition, the model is similar to Hansen's and does not appear to be any more accurate (Swerdloff and Stowers, 1966). In fact, when Swerdloff and Stowers made a comparative evaluation of the performance of the Hansen, Stouffer and Schneider models they concluded that none of the models was very accurate although the Hansen version gave marginally better forecasts in their tests. But they also added that differences between the models were not large enough to warrant recommendation of any single one in preference to the others. Harris (1966a) has since proposed another form of accessibility model that is quite different from Schneider's, in which trips fall off over opportunities rather than distance. This has, however, not yet been tested.

Other versions of the gravity model have been used in retail analysis. The best known of these are the Lakschmanan and Hansen (1965a) retail model, the Huff model (1963; 1965) and the Rogers model (1965). However the formulations are very similar to Hansen's accessibility model. They state that the sales at a shopping centre are directly related to the size (and sometimes quality) of the centre and inversely related to distance from consumers. But applications of these models to Baltimore (Lakschmanan and Hansen, 1965b) and to Lewisham in London (Rhodes and Whitaker, 1967) have indicated that they are not very accurate and are not particularly useful except perhaps for evaluating plans at the large shopping centre scale. In its present form the models have not been designed for accurate forecasting.

Quite apart from these operational limitations, the gravity model has several other drawbacks (see Huff, 1965; Carrothers, 1956). Firstly, as Lowry (1964) argues, the model has rather weak behavioural underpinnings, and hence has low theoretical content. The model is essentially descriptive; it describes aggregate interaction patterns but offers no explanation of them. This is because the model does not express a theory of rational consumer choice (Lowry, 1964, 22). Consequently the parameters will vary with the size and composition of the opportunities. For example, the distance exponent (b in equation 2) varies by socio-economic group, mode of travel (Alcaly, 1967) and size of city (Voorhees and Shofer, 1966). There is also no reason to assume that it does not also vary by time of day, week and over a period of years. Furthermore it has been suggested by Carrothers (1956) that the distance exponent itself may be a variable that is inversely related to distance. Finally, the model is sensitive to the interpretation and measurement of the attraction variable (see Catton, 1965) For instance, in a retail model, different results will be obtained from using number of employees, square feet of selling space, turnover, or any of these raised to some power, as measures of the attraction of a shopping centre.

Although the gravity model is therefore an *a priori* model, it has by no means been abandoned in urban models. Most of the larger recent residential models use the concept as a measure of accessibility in one or more parts of the model (see for example C. C. Harris, 1966; Schlager, 1964; Ellis, 1966), but the concept no longer dominates residential model designs. On the other hand in retail models, where interaction rather than choice of residential location is being modelled, the gravity formulation is still very important. Roger's model (1965) for the Bay Area Transportation Study, Graybeal's retail development model (1967a), Huff's probabilistic model (1965; 1966) and the Voorhees model for Canberra (1966), as well as most of the proposed retail models for British cities and regional shopping centre systems, resemble closely the Lakschmanan–Hansen formulation, although sometimes they introduce additional disaggregation of shopping trips and retail activities.

3 Contributions from geography

Central place theory has received considerable attention from geographers and it appears to have some relevance to intra-urban analysis. The concepts as originally derived by Christaller and Lösch were postulated for a system of cities, but some of the ideas have been tested or examined in intra-urban context. The real issue to be explored here is whether central place ideas have produced either useful conceptual or predictive models of city growth and structure.

Actual testing of the theory within a city is clearly not possible given the simplifying assumptions of the original model, but central place ideas have sparked off a good deal of research into the internal structure of cities (Berry, Barnum *et al.*, 1962; Berry, 1963; Simmons, 1966; Garner, 1967). This research has not validated the original theory but it has identified hierarchies of centres within cities, although these are not so clearly defined, complete or regular as might be suggested by theory. Similarly there is no strong evidence available that the centre structure of cities is stepped in a regular manner.

The only model of city structure that has so far arisen from this research is Berry's retail model (Berry, 1963; 1965) for Chicago. This expresses the relationships between the number of establishments in a trade area and the population and income of the trade area. These equations do not actually operationalise central place theory but they do give some support to the idea that centre size is related to the spending power of trade areas.

However the model has some limitations as a predictive device. It requires that trade areas can be established and defined in the city. This can be a very difficult problem given the hierarchical nature of trade areas and the degree of overlap, and even if they can be defined for model calibration, predicting the size and shape of these areas for the forecast period would be extremely difficult. Other retail models, such as the Huff and Lakschmanan–Hansen models, do not make this limiting assumption about trade areas and thus permit an unlimited degree of overlap and interaction.

Berry's model is not therefore a very good predictive model nor is it a conceptual model that contributes a great deal to our understanding of the structure of the city. In order to increase our understanding a more useful approach might be that followed by Golledge and Brown (1967) on the stochastic learning and search behaviour of shoppers, or by Downs (1969) on the consumer perception of shopping centres. These behavioural approaches to retail analysis and urban structure may lead ultimately to a reformulation of central place concepts under different assumptions, and in a way that will make them applicable to intra-urban retail location.

4 The contributions of urban economists

Urban economics has probably had a greater impact on the formulation of theories and models of urban processes than any other single discipline. Based on the tenets of classical price and rent theory, the theory of urban land use as expressed by Haig (1925), Ratcliffe (1949), Wendt (1957) and others, states that sites in the urban area are bid for in the land market, and are occupied, after competitive bidding, by the best use, i.e. that use which is prepared to pay most to purchase the bundle of attributes and qualities that constitute that particular site.

The theory therefore proposes that it is possible, under the assumption of perfect competition, to attain a position of market equilibrium in land where each activity finds its own rent level in such a way that the aggregate rent-paying ability of the whole system is maximised. The shape and topography of the resulting rent surface is assumed to be a function of the utility of sites to different activities generally measured in models by accessibility to work, shops, markets and different types of urban environments. Each site has in effect a unique combination of characteristics associated with particular land uses, and the use that pays most for its own desired group of attributes occupies the site.

This concept has been developed in models of residential location by Alonso (1960) and Wingo (1961). They assume a single employment centre, a socially homogeneous population, and a perfectly symmetrically network of transport routes, thus generating simple rent surfaces with the basic locational criterion of the minimisation of transport costs to the employment centre.

Although these formulations make simplifying assumptions that preclude operationalisation of the models, they do state the urban growth problem in terms of the actors or decision makers in the system. In the case of Alonso and Wingo, the actors are household units whose decision to locate is dependent upon the household budget, transport costs and location of the workplace. This is in several ways unrealistic because there are many other factors, some possibly of more importance, affecting a household's choice of residential location (Chapin and Weiss, 1962; Lansing and Barth, 1964). Moreover simplification of urban spatial structure into a homogeneous residential area with uniformly good services and one employment centre is also unrealistic. Yet the critical importance of the micro-economic approach as expressed in these early models was the emphasis on the *process* of change in urban land use.

Despite the conceptual advantages of this approach, many land use models have no behavioural or market interpretation. In fact the operational models which have been used for forecasting have least in common with the work of the urban economists. Lowry (1967) in his recent review of urban models looks at five important models, the Chicago Area Transportation Study Allocation Model, the Chapin and Weiss Model, the EMPIRIC and POLIMETRIC models developed for Boston, and Lowry's own Pittsburgh model, none of which has much economic interpretation because none considers how and why locational decisions are made. In these models urban growth is a function of the potential of sites for development, not of competition and transactions in the urban land market.

Lowry is more impressed with models that specify the operation of the 'market-clearing process', by which households, firms and shops are assigned to sites in accordance with their respective budgets. Models like the Herbert–Stevens model (1960) for Penn Jersey, and the San Francisco Housing Market Model (Robinson, 1965; Little, 1966) do attempt to simulate this process. The Herbert–Stevens model and later extensions and reformulations of it (Harris, 1966b) use a linear programming structure to maximise, in the primal case, the aggregate rent-paying ability of all new households located in the urban area. This amounts to both maximisation of household budget savings (and at the same time landlord rental income, by the definition of the market equilibrium), and minimisation of transport costs in the entire region. Thus new households are distributed over the city in a configuration that is optimal from the point of view of all households to be allocated. This configuration, as Stevens points out, is optimal in a Pareto sense; no household can move to increase its savings without reducing the savings of some other household and simultaneously reducing aggregate savings.

One of the difficulties of models of this type arises when attempts are made to interpret the solution in terms of the real behaviour of the land market. While the model might be able to simulate the pattern of households responding to a given rent surface, the key issue is how bid rents are established in the real world. Clearly access to work is only one dimension of the problem. Thus Harris has been working on defining and measuring the preference structure of different household groups with the eventual objective of developing utility functions for these groups (Harris, 1966c). But the measurement of these functions requires an estimation of the value which persons place on their expenditures for different commodities, and the reasons for the variations in patterns of preference within an urban population. This is so obviously complex and related to psychological variables, personal aspirations and non-economic behaviour, that the adoption of these ideas in operational models cannot yet be foreseen.

The San Francisco housing model is a similar market model to the Penn Jersey with many of the behavioural determinants of supply and demand not specified explicitly. However the model is not an optimising model, but one that seeks a market-clearing solution by making adjustments in housing stock as pressures of demand fluctuate over time, yet never actually attains this solution because demand is never stable and existing stock deteriorates over time. Disaggregation of supply units, i.e. household types and demand, is much greater than in the Penn Jersey residential model, and hence it is a more precise simulation of market transactions. It is also a more refined simulation with the ability to trace the effects of demand pressure through the market from changes in rent levels to the responses of builders to opportunities to construct housing units if this appears to be profitable. The model then is a large computer program containing a large number of instructions and decision rules that is continually matching housing groups with house types and rent levels. It is essentially similar to Penn Jersey in its assumption that households seek groups of housing type attributes, but it does not produce an equilibrium solution.

As Lowry points out (1966), a further characteristic of the model is that it does not assign households directly to locations in the city. The model is aspatial in character in that households are deterministically assigned to matching housing categories wherever they might be located. This is a serious disadvantage because households are not able to discriminate at all between different social areas, nor to take into account accessibilities of any kind in their locational decisions; nor indeed, to choose a *sub-optimum* housing unit or location. This kind of deterministic matching is thus, as Ellis has pointed out (Ellis, 1966, 113), not the most realistic way of simulating the way in which potential household locators search through both housing type *and* locational files during the decision process. The San Francisco model is not, therefore, a locational model in the same way as the Herbert–Stevens model nor in the Alonso or Wingo sense, although it does not attempt a market interpretation of the urban growth process.

5 The divergence of theoretical models and predictive models

It will be apparent from the discussion above that there is considerable variation in the theoretical and behavioural content of urban models. Generally it appears that the more a model is committed to theoretical concepts and simulation of behaviour of decision makers in the urban system the more difficult it is to operationalise and to use for forecasting. Clearly predictive models are not explanatory models, and as Blalock points out (1961, 10) it would be misleading to confuse causal notions with those of prediction.

However this divergence can be explained by the fact that most of the early land use models were produced almost to order to provide the necessary land use forecast inputs to traffic generation and assignment models, while research on explanatory models, which were built with different objectives, continued almost independently until the nineteen sixties. These two streams of interest are schematically shown in Figure 1.

The first generation models were sponsored by United States government departments and city transportation authorities as part of the transportation planning process (United States Department of Commerce, 1962). The function of these models was to make accurate land-use forecasts for traffic zones or districts in the city, and not explicitly to explore or to express the processes of land use change. The Chicago Area

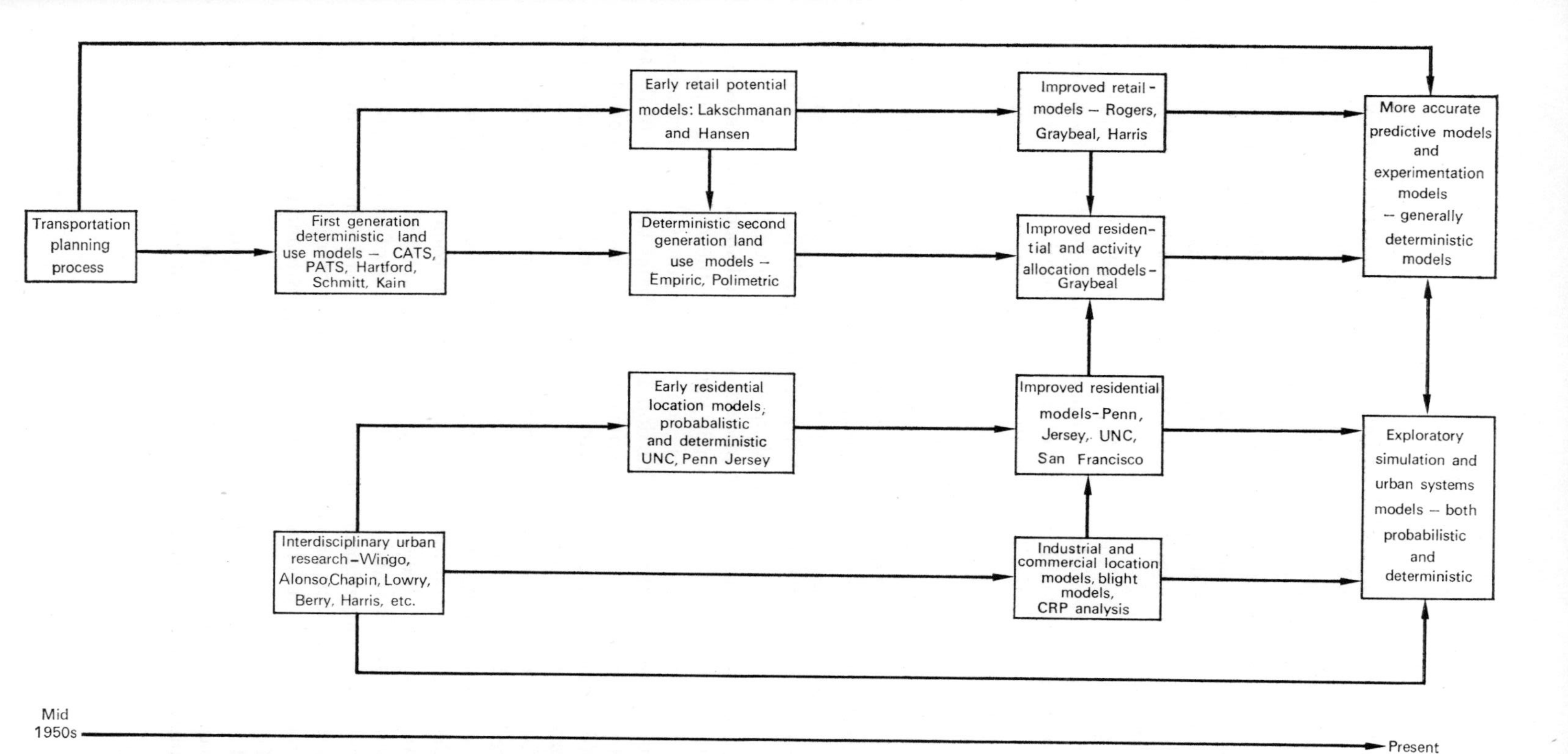

Fig. 1 Sequence of developments in urban modelling since the 1950s.

Transportation Study (CATS) land use model (Hamburg and Sharkey, 1961) was followed by models for Detroit (Kain, 1962), Hartford (Voorhees and Associates, 1962), Pittsburgh (Lowry, 1964) and other cities (see also Schmitt, 1954; Hamburg, 1959; Hansen, 1961; Voorhees, 1961; Hansen, W. B., 1961; Lakschmanan, 1964, for similar types of models). The models were not explanatory (although they were not always good predictive tools either), but it would be inaccurate to suggest that they were simply statistical with no theoretical underpinnings. Concepts that underlaid the models were loosely defined and behavioural elements treated indirectly.

A good example of the difficulties encountered when an attempt is made to incorporate theoretical ideas in a predictive model can be found in Kain's multiple equation model of residential location calibrated to data from the Detroit Area Transportation Study (Kain, 1962). His household equation is—

$$R_{ij} = f(F_{ij}, Y_{ij}, P_j, S_{ij}, N_{ij}) \tag{3}$$

where R_{ij} is the residential space consumption of the ith worker at the jth workplace, P_j is a proxy variable for the price of residential space at the jth workplace; S_{ij} is the sex of the ith worker at the jth workplace; N_{ij} is the labour force participation by ij's family, i.e. number of employed persons in the family; F_{ij} is the size of ith worker's family at j. The most important theoretical gap in this model, as Kain well appreciates, is the use of a proxy variable for the price of residential space: this turns out to be easily the most powerful single variable in the estimating equation. The proxy used is the distance of each zone from the CBD, which assumes that a linear distance decay function approximates the shape of the bid-rent surface of the city. This is quite clearly an over-simplification of a key decision process in urban growth, and is the inevitable product of inappropriate data resources.

Usually model builders, constrained by time and data limitations, are forced to fall back on such surrogates for important processes and theoretical concepts. Accessibility functions describe the frictional effect of distance on social and economic interaction; zonal attraction indices represent a large bundle of residential site characteristics; and journey to work distances stand in for rent surfaces. Some of this generalisation is a result of the aggregation of processes and activities in space and time, but it partly reflects also a preoccupation with predictive models, rather than models which actually simulate urban growth and explore or express urban theory.

This is quite understandable, but means that many of the so-called urban models are simply extrapolations of statistically significant regularities. Thus a model of residential growth might look like:

$$\text{Growth of Population}_i = f(\text{Access to Employment}_i, \text{Availability of Sewer and Water Services}_i, \text{Quality of Transit Service}_i, \text{Location of Schools}_i, \text{Median Income}_i) \quad (4)$$

This kind of model, neglecting for a moment the interdependencies in the equation, might be an excellent predictor of zonal population growth, but it tells us nothing about:

1 Levels of in- and out-migration associated with each zone
2 Rate of population growth within the zone
3 Origins of in-migrants
4 Destinations of out-migrants
5 Type of population or housing in each zone
6 Reasons for in- and out-migration, internal growth, or internal decline.

It may be adequate for traffic assignment models, but does not provide many insights into urban development processes. Simulation of these processes in a predictive model may, however, be prohibitively complex, and to illustrate this point two models built for Boston are compared. The first, EMPIRIC, is a straightforward forecasting model, and the second, POLIMETRIC, is a conceptual improvement to that model.

The EMPIRIC model designed by Traffic Research Corporation (Hill, 1965; 1966) is a good example of what Lowry (1967) calls a 'row model'. A row model is concerned with changing locational patterns without considering the succession of land uses that may be responsible for these patterns (a 'column model'). The model is made up of a system of equations that forecast the distribution of population and employment as a function of a set of *locator* variables that influence the location of activities, and a set of *located* variables that have to be allocated endogenously by the model. The locator variables are zoning practices, auto and transit accessibilities. land use intensities, and quality of water and sewage service; while the located variables are white and blue collar workers, retail and wholesale employment, manufacturing and other employment.

The model states that a change in the subregional share of a located variable in each subregion is proportional to the change in subregional share of all other located variables, and the change in the subregional share of all locator variables in the subregion. Mathematically, the model is expressed as:

$$dR_i = \sum_{=1}^{n} a_{ij}\, dR_j + \sum_{k=1}^{m} b_{ik}\, dZ_k \quad (5)$$

where i or $j = 1, 2, i, j, \ldots, n$ number of located variables; $k = 1, 2, k, \ldots, m$ number of locator variables; dR_i or j is the change in the level of the ith or jth located variable over the forecast period; dZ_k is the change in the level of the kth locator variable over the forecast period; and a_{ij}, b_{jk} are parameters. The model was calibrated to 1950–60 time-series data and obtained a very close fit indeed with the actual data with very high R^2 values for all dependent variables and correspondingly low mean square error ratios. However the model gives few insights into how the subregional growth that it describes takes place. The spatial patterns of intra-urban migration are not specified, nor is the succession of uses on sites in the city. A further factor is that change is described by sets of statistical regularities rather than as the product of decision by locators.

Because of some of these limitations, another version of the model known as POLIMETRIC was developed by Traffic Research Corporation (TRC, 1963). This model specified the interzonal flows of persons and activities, but was not calibrated so successfully as EMPIRIC. The general form of the model is:

$$dR_i(e) = p(e)\, R_i(e) + \sum_{j=1} M_{ji}(e) - \sum_{j=1} M_{ij}(e) \tag{6}$$

where $R_i(e)$ is the value of activity e in subregion i; $p(e)$ is the percentage subregional growth of activity e during forecast period;

$$\sum_{j=1}^{n} dM_{ji}(e)$$

is the total amount of activity e which migrates from all other subregions j to subregion i during the forecast period;

$$\sum_{j=1}^{n} dM_{ij}(e)$$

is the total amount of activity e which migrates to all other subregions j from subregion i during the forecast period; and where n is the number of subregions, and m the number of activities. The model therefore calculates change from the difference between in- and out-migration. But in doing so it is necessary to compute migration functions for each activity—i.e. propensity to move factors—and also measures of the desirability of each subregion for each activity. This was done, but the model proved to be computationally cumbersome and did not give such a good fit as EMPIRIC. It should be noted that the model is still a row model in the Lowry sense, but keeps explicit accounts of changes in activity levels. It also illustrates the difficulty of increasing the behavioural content of the model without obtaining correspondingly better fits to data. Moreover the model requires special data sets to test particular hypotheses, and cannot be run with data collected for other purposes (such as in origin–destination surveys).

Consequently, as the model becomes an increasingly realistic simulator of behaviour in the city (as for example, the University of North Carolina, Penn Jersey and San Francisco models) more complex data is required and calibration and accurate prediction become more difficult and expensive. One further important point is that, even if the model is able to replicate some actual distribution of activities in the city, it cannot be used for predictive purposes unless the same data used for calibration is available for the forecast. This would not matter if the model was what Schlager (1964) calls a 'design' model whose function is to evaluate alternate plans rather than make accurate predictions for input to a traffic assignment model. Thus, the model builder has to make trade-offs between complexity and forecasting ability: decisions that are circumscribed by time, data, resources and objectives. These decisions are also dependent upon other choices the model builder must make. In the next section the aggregation problem is examined in this context.

III Choice of the level of spatial aggregation

1 Grouping activities and areal units

It is inevitable in a mathematical model, or in any other kind of abstraction from the real world, that some aggregation of activities, areal units or decision units must take place. Since these decisions affect the level of spatial detail that the model builder can work at, the problem is considered here as a general one of spatial aggregation.

The problem is, in a sense, one of classification or discrimination between groups for a particular purpose. The objective is to find a set of activities or areal units that have minimum within-class variance and maximum between-class variance. We are looking for classes that are homogeneous and can thus be treated in the model as single entities. But this process is only useful if the classes produced are meaningful units of analysis or behaviour.

A great variety of levels of aggregation has been adopted in land-use models, and a number of different methods of grouping has been used. Generally, it seems that the more aggregated activities and areal units are, the more simple the model, and the greater the likelihood that the model is a forecasting rather than a theoretical model. For example, at one extreme the CATS model (Hamburg and Sharkey, 1961) and EMPIRIC (Hill, 1965) distribute fairly broad categories of population and employment; while at the other extreme, the San Francisco model has 22 categories of dwelling unit, 14 different types of residential location and 4,980 tracts or homogeneous neighbourhood segments. Usually, the selection of the categories in a model depends on the way data is collected in origin–destination surveys or in published sources such as the census. The population projections for the Baltimore Retail Model (Lakschmanan and Hansen, 1965) were based on aggregated time-series data from the 1950 Census and the 1960 Baltimore Area Transportation Study estimates. In contrast, Ellis (1966) uses multiple discriminant analysis to determine the functional groupings of households for each of his environmental area groups. He suggests also (Ellis, 1966, 115) that the method might be profitably applied to all variables, age of housing, racial composition of an area, and social status of the area, thus breaking variables down into discrete groups rather than using the more traditional continuous measurement scales. The aggregation problem is, therefore, first a measurement problem at the data collection level, and then a grouping problem if information is considered significantly heterogeneous.

However in some of the more recent models (see Chapin and Weiss, 1965; Orcutt, Greenburger *et al.*, 1961) complete disaggregation has been the objective, or at least the theoretical starting point for analysis. Complete disaggregation means focusing on the individual decision makers or decision groups in the urban system and hence working at a very fine level of spatial detail. Rogers (1965, 70) in his review of retail land use models states that: '. . . any sound theoretical attempt to interpret retail spatial structure must begin with the consumer.' And this approach has always been the viewpoint of Britton Harris with regard to Penn Jersey (1961), and of Chapin and his associates at Chapel Hill (1962).

Yet the difficulty of working at this level has encouraged most workers to aggregate the decision units to some extent. In fact, it might be argued that the behaviour and interaction of *individual decision makers* might be more suitably analysed in land use gaming models (Duke and Schmidt, 1965), while aggregate behaviour patterns could be investigated in urban growth models. This does, of course, necessitate grouping together similar behaviour units such as households with similar locational requirements, or consumers with the same shopping habits. This is the approach adopted in practice by Donnelly, Chapin and others (1964) in the University of North Carolina model, by Rogers (1965) in his proposed retail model, and by Graybeal (1967b) in his land use simulation model for the Bay Area Transportation Study.

2 Choice of scale and spatial autocorrelation

The level of spatial aggregation selected in a model—i.e. traffic zone, census tract, enumeration unit or grid square—determines the scale a model can work at. But because there is a close relationship between scale and the identification and measurement of spatial series (recurring spatial patterns), this choice is an important one. Tobler (1966) and Harvey (1968) discuss the autocorrelation problem and point out that the size and shape of areal units determine the ranges of spatial frequencies that can be detected in a spatial series. Consequently, as the level of aggregation increases, more and more spatial frequencies may be obscured, therefore disguising important processes in the urban system (also see Cliff, 1968).

3 Errors in predictions caused by aggregation errors

A more familiar problem of aggregation is the variation of results from a model which is applied to areal units at different levels of aggregation (Duncan, Cuzzort and Duncan, 1961). If a set of traffic zones, for example, were used as the observation set in the calibration of a model, but were later aggregated to traffic districts for a further calibration, it is unlikely that the coefficients in the two models would be similar. The difference would be partly due to the scale problem mentioned above, and partly due to the different autocorrelation values for the two sets of areal unit.

Two recent examples illustrate some of the errors that the model builder can encounter. One of the reasons for the unsuccessful performance of the POLIMETRIC model was the unequal distribution of activity levels through all the subregions. Boston was included in one subregion although it contained 30% of the regional population and 40% of the employment. This meant that the distribution of activities was very skewed, and owing to the disproportionate weight of the larger subregions, the model was actually fitting the coefficients to 10 or fewer data points per activity when there were 29 subregions (TRC, 1963, 26).

In the second example, Horton (1967), using the CATS data, tested the hypothesis that there was a significant difference in the accuracy of travel forecasts at different levels of land use type and areal aggregations. Firstly, he discovered that there were significant differences between person trips and the square feet of land use for each *subgroup* of the major land use categories employed by CATS. Secondly, this relationship proved to be unstable over different areal aggregations. Calculating coefficients for observations in six concentric rings drawn around the CBD, he found highly significant differences in the values of coefficients.

Hence, there is enough evidence available to cast serious doubt on the validity of models that are calibrated at one areal level, say the traffic district, and which are then used for forecasting zone activity levels. The converse of this procedure, in which forecasts are made from small to large-scale units does not, intuitively, appear so dangerous. But McCarty (1956, 263) concludes his discussion of scale by stating that:

> ... in geographical investigation it is apparent that conclusions derived from studies made at one scale should not be expected to apply to problems whose data are expressed at other scales.

IV Choice of the level of time aggregation

1 Cross-sectional versus change models

The time component of the model is generally prescribed by data availability in the same way as the choice of areal aggregation. The dynamic nature of the urban system suggests the use of change, rather than cross-sectional models (i.e. models that are calibrated to one point in time) should be used where possible, and if this is not feasible, then inferences from cross-sectional models should be made with great care.

There are considerable dangers in assuming that relationships (parameters) established in a cross-sectional model will repeat themselves in a change formulation of the same model (Duncan, Cuzzort and Duncan, 1961, 121). For instance, a close relationship might be established between gross residential density and income at one point in time. However if there were an increasing number of high and middle income persons moving into apartments because of space restrictions (as in a city like Caracas, for example) or a change in living tastes, it would be very dangerous to infer stability of this parameter over time. Hence, where the two types of model are likely to give contrary results, cross-sectional parameters will not make good predictors.

A further disadvantage of the cross-sectional approach is that it generally implies simulation of what Lowry has termed 'an instant city' (1964, 39). Lowry's Pittsburgh model only implies a time dimension in the iterative sequences of finding a solution, but does not allocate activities to an existing surface of activities. Each new pattern of urban development is predicted independently which for predictive purposes could be very risky.

'Ideally, a model of metropolis should be a dynamic system' (Lowry, 1964, 40). The model should incorporate the time lags, inertia factors, sequences of decisions and dynamism that are apparent in any city system. Change models using time-series data for two points in time are a step in this direction. The models may be made up of sets of differential equations describing the rates of change in activities over the calibration period (Hill, 1965; Scott, 1968) or they may be dynamic programming models that generate some optimum solution every five years as Ellis proposes in his planning model design (1966). In other words the models simulate additions to the existing urban pattern and are not responsible for simulating instantly the present-day resultant of an entire history of city growth. A change model can at least assume stability of parameters over its calibration period, but a cross-sectional model uses relationships that appear reasonable from the point of view of present-day urban growth for simulating across a period in which these relationships could not be expected to hold.

2 Measuring change

One of the problems the model builder has to face is choice of measurement of change. Change, as Duncan, Cuzzort and Duncan (1961) describe, can be expressed in a number of ways. These include using absolute, relative or proportional change, and deviational change. All of these have been used from time to time in urban models, but more complex measures (such as differential shift) have been in a number of activity allocation models (Voorhees and Associates, 1966; Litton Industries, 1965). The shift methodology states that change in the activity level of a subregion is related both to the proportional share of total regional growth which would accrue to the subregion if it grew at the same rate as the region as a whole, and to the share due to the differential growth rates of region and subregion for a particular activity.

The shift approach is really no more than a convenient device for allocating regional totals to subregions, and it does not have any strong conceptual justification. While the model uses time-series data, it does not simulate the changes in such a way that gives any indication of how or why the changes take place. Because of this, the differential growth component of the measure is often calculated endogeneously within the model as a function of a number of specific locational variables (Voorhees and Associates, 1966), and not in the manner described above.

3 The problem of autocorrelation

In the same way that autocorrelation can occur in a spatial pattern—i.e. observation points are not mutually independent—the values of variables at different points in time are almost always related because some time trend is operative. We can specify this relationship to time in a variety of

Serial correlation

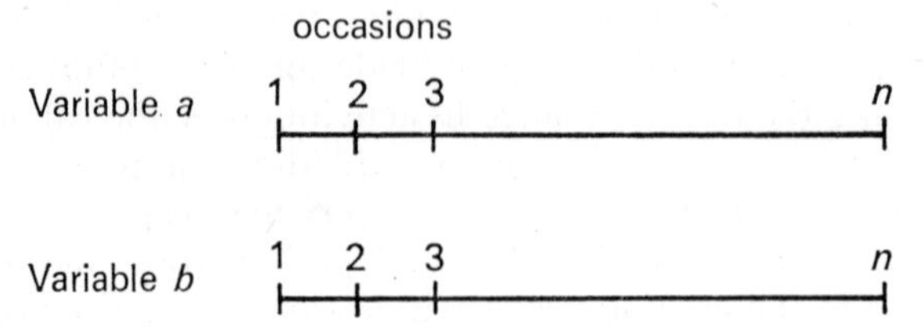

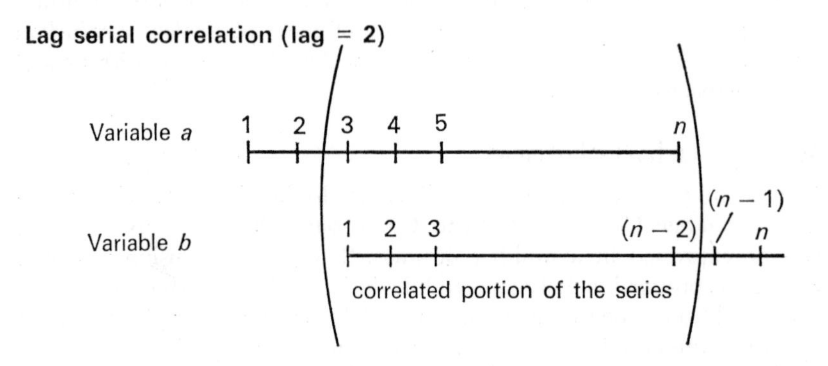

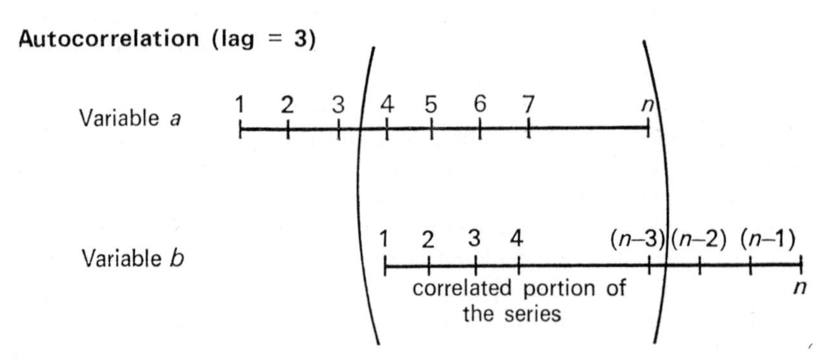

Fig. 2 Serial and autocorrelation.

ways: the series might be linearly related (as in a linear model) or it might be periodic (as in a sinusoidal function). The model builder's problem is to specify correctly this function so that autocorrelation is accounted for and predictions over time made accurately.

The problem may also express itself in the technical issue of avoiding autocorrelated error terms (the residual variance) in a linear model such as a multiple regression equation. Here, one of the assumptions of the model is that the error term e, in a model like equation (7) which incorporates lag variables, is independent of all past, present and future values of the independent variable X (Johnston, 1963, 207).

$$Y_t = a + bX_{t-1} + e_t \tag{7}$$

However it is equally important to recognise autocorrelation operating within the change model itself. Cattell (1966, 382) distinguishes between serial correlation, lag serial correlation and autocorrelation. Serial correlation he defines as correlation among different time co-ordinated variables, whilst true autocorrelation is the lagged correlation of one variable measured over time with *itself*. Cattell illustrates these as in Figure 2. Cattell further makes the important point that smoothing out the cycles, trends, frequencies in time-series data is not a real solution. The level of aggregation or degree of smoothing is critical to our understanding of urban processes. Chapin's recent work (1967) is particularly relevant here. He is attempting to identify the temporal routines and cyclic behaviour of key decision groups in the city. Harris (1961, 703) has already stressed '. . . the overwhelming importance of the diurnal cycle of family living and participation in the production process.' Chapin is investigating also other decision cycles that may have distinct behavioural patterns, such as migration, business relocation and recreation. Empirical investigation into these problems has been started by Chapin and Hightower (1966) and by Cowie (1968) who is concerned with intra-urban migration and the decision to move.

It is therefore essential to our understanding of urban spatial processes to recognise time and space correlated variance. Cattell (1966, 384) comments that '. . . the methods used by economists of partialling out completely any time trend is not the answer, and is as theoretically vicious as it is widespread.' And he concludes by wondering '. . . how much of significance would be left in life if we removed everything that has been important in terms of trend and historical change?'

V Choice of model structure

The model structure is the mathematical framework or statistical technique that expresses the relationships between variables and the rules for modelling a certain concept. Model builders have tried a variety of methods—multiple regression and simultaneous equation systems (Lowry, 1964; Hill, 1965; Kain, 1962; Scott, 1968); linear programming (Herbert and Stevens, 1960; Schlager, 1964); and probabilistic models such as Markov chains (C. C. Harris, 1966), and Monte Carlo simulation (Morrill, 1965); Chapin and Weiss, 1964). This range has been discussed fully by Harris (1967), so that attention will be restricted to linear models and to Monte Carlo simulation models. On the one hand, linear models are the most frequently used and criticised, and on the other, Monte Carlo simulation is often offered as the desirable alternative since it has few of the disadvantages that the former possess. Moreover, linear models are representative of the deterministic modelling approach while Monte Carlo simulation is probabilistic.

1 Linear predictive models

Most of the spatial or activity allocation and market potential models (Lakschmanan and Hansen, 1965) in the literature are linear models. This means that they are made up of systems of linear equations and assume that the relationships between the variables are linear—i.e. that the coefficients are linear estimators. It also means that the models are committed to a mathematical framework that has been devised for analysis of *experimental* rather than *observational* data (Wold, 1956). Linear models have, therefore, been devised for laboratory-controlled experiments and not social science problems. The result is that the model builder must attempt to force his rather clumsy experiment into a highly restrictive analytical framework. Some of the difficulties of the framework will be examined particularly in relation to the types of error likely to be incurred during the model building process.

Two major categories of error are encountered when any kind of model is formulated. The first is *measurement error* which includes data collection, sampling, and scaling errors, and is related to the experimental design. The second category, and one the model builder is more likely to be able to control for, is *specification* error which arises from the manner in which the model is specified and varies with the degree to which the model meets the mathematical assumptions of the technique.

a Measurement errors: These are discussed at length in the literature on measurement (Ellis, 1966), sampling (Cochran, 1963; Moser, 1958; Stuart, 1958) and scaling (Torgerson, 1965). All these aspects of designing experiments, collecting data for models, using the right measures and measurement scales, are integral parts of model building and become of increasing importance as special data sets are collected for models that incorporate behavioural variables and operate with highly disaggregated data. It should be emphasised that the predictive accuracy of a model, and the inferences that can be drawn from it, are as much dependent on the level of measurement error as on the specification errors looked at in detail below.

b Specification errors: The linear model makes a number of assumptions that have to be met if the model is to be properly specified. Failure to meet or recognise these assumptions leads to errors that can produce erroneous predictions and spurious inferences (Wold, 1956).

The first assumption is that of normality. We must assume that the input data for the model is normally distributed (and incidentally, that the distribution of residual variance, as in equation (7), is also normal). However symmetrical distributions are practically non-existent in social and behavioural data (Ferber and Verdoorn, 1962, 64). For example, distributions of income, journey to work trips, household expenditures, and city size are not normal. Other functions—such as the Pareto, log-normal or other gamma functions—can best be fitted to these distributions. A Pareto function can be fitted to the distribution of New York daily newspaper firms (Cohen, 1966), while Morrill (1963) describes a set of gamma functions that can be fitted to migration data.

Consequently, transformations are necessary to transform data into a suitable form for use in linear models (Bartlett, 1947; Box and Tidwell, 1962). Generally, logarithmic, trigonometric and exponential transformations are made if the distributions are monotonic, and sine curves or *S*-curves if they are not.

One of the consequences of introducing non-normal data or incurring non-normal error terms in a linear model, is that it becomes difficult to interpret the usual measures of the mean, standard deviation, and variance. Furthermore, parametric tests such as *t*- and *F*-tests and other tests of significance are affected. Not a great deal of work has been done on the affect of non-normality on the power of tests although Pearson (1931) suggests that analysis of variance and *F*-tests have reduced power. But other writers conclude that tests are fairly insensitive to non-normal errors (Malinvaud, 1966, 252). Box and Watson (1962), however, have shown that if the theoretical degrees of freedom are increased by some factor, the tests can be applied satisfactorily. Quite often in practice the model builder will eliminate extreme observational values that are likely to distort distributions either before the data are submitted to the model, or afterwards when results appear to be unreasonable.

The second assumption of the linear model is that the error terms are homoscedastic, that is they have homogeneous normal variance. This will not be too serious if the error terms in a system of equations are mutually independent, but the linear estimators will become biased if the variances do diverge and become heteroscedastic. The model will then be highly sensitive and the power of parametric tests will be affected (Malinvaud, 1966, 256).

A third assumption of the model is that there is linearity of relationships with respect to the parameters. The coefficients are therefore assumed to be linear estimators, expressing linear relationships between dependent and independent variables. This linearity concept is disarmingly simple and this is its virtue, but as Horst (1966, 134) points out, this may be gratuitous in view of the arbitrary scaling procedures adopted in most multivariate analysis techniques. Digman (1966, 465) makes an interesting suggestion that non-linearity is most likely when we examine relationships between what he calls domains.

A domain is a characteristic or dimension of some phenomena that can be measured in a number of ways (in psychology, by a battery of tests for example). Thus, intelligence, physique, and visual perception are domains within which each test or measure on a variable would be linearly related. But when relations between domains are examined then non-linearities are likely. To take an example from urban analysis, residential environmental quality might be measured by a number of variables, but relationships between this factor (in factor analytic terms) and another factor such as social class might be non-linear.

However there are many other reasons for non-linearity and there is considerable difficulty in deciding whether a relationship is truly non-linear or the result of poor measurement or scaling, as Horst (1966) suggests. When this mistake is made, spurious inferences and inaccurate predictions will follow.

If it can be determined that non-linearity is present when variables are plotted against each other, there are a number of methods for dealing with it so that we can apply the linear model. Generally, the relationship is transformed. Thus:

$$Y = kX^bZ^c \tag{8}$$

becomes, after transformation,

$$\text{Log } Y = \text{Log } k + b(\text{Log } X) + c(\text{Log } Z) \tag{9}$$

Similarly, semi-log, reciprocal and other transformations can be made to straighten out relationships so that a linear parameter can be fitted. Alternatively, polynomials might be fitted to a curve provided that each additional higher order exponential term added to the function is judged significant (Larson and Bancroft, 1963). A third approach, though less common, is division of the population into sub-samples by dividing the curve into several linear sections. As long as the intervals are meaningful in terms of the problem under investigation, this is quite justifiable. A further alternative is to ignore the non-linearity altogether. This would be a suitable solution where the model is designed for short-term predictions and relationships are not changing rapidly over time.

Another restricting assumption of the linear framework is that variables on the right hand side of a linear equation are independent of each other. In other words, the independent variables should not be intercorrelated, i.e. they should have correlation coefficients of zero. However interdependencies in linear models are unavoidable because of the complex interrelationships in an urban system, as equations (4) and (5) demonstrate. The multicollinearity can under some circumstances be ignored if the same structure of interrelations can be expected to continue into the future (Johnston, 1963, 207), but if this cannot be guaranteed then exploration of the structure is necessary.

However there is no standard procedure for dealing with multicollinearity, although the first step is usually to examine the matrices of simple and partial correlations between all variables before making a commitment to any group of variables for the model. If two variables are found to be highly intercorrelated one can be dropped from the model, because a linear combination of two closely related variables gives the same result as just one variable. Ideally it would be desirable to subject all variables to a factor analysis to break out the underlying dimensions of the data set, but often there are too few variables in an urban growth model to make this worthwhile. And even if this procedure is carried out, it remains doubtful whether these factors should be used for prediction (Matalas and Reiher, 1967).

Yet it should be stressed that the inclusion or exclusion of variables from a linear model can make a great deal of difference to the values and signs of the regression coefficients, which are, after all the predictors in a model. Wallis (1965) has demonstrated this using known relationships between the dimensions of a cylinder. Clearly, if the linear model is sensitive to variable selection and intercorrelation, great care should be taken in choosing variables.

The final assumption to be dealt with here is the additivity assumption. It is assumed that the error terms are neither multiplicative nor related in any way to the independent variables, but are simply additive with small variance. If the residual variance is small, a high correlation between the error terms and the independent variables will not affect the value of the parameters. Similarly a low correlation will not affect parameter estimation.

Accumulation of errors in linear models: Both measurement and specification errors in a model, and others that may be hidden in the exogeneous variables, accumulate during the model building process and are expressed finally in the residual variance of the model. But if a model or equation forms a sub-model in a larger model system then errors become compounded as models are merged and fed into one another. This is particularly likely in a recursive chain type system (Alonso, 1967) such as in equation (10)

$$\begin{aligned} Y_t &= a + bX_t + e_t \\ Y_{t+1} &= a + bX_{t+1} + Y_t + e_{t+1} \\ Y_{t+2} &= a + bX_{t+2} + Y_{t+1} + e_{t+2} \end{aligned} \tag{10}$$

Alonso (1967) cites an example of a three-step regression chain similar to the one above each with an $R = 0{\cdot}9$. He calculates that the result of such

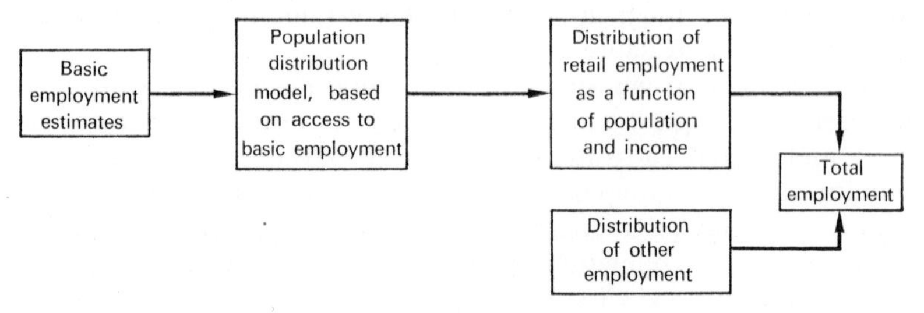

Fig. 3 The chain structure of a typical land use model.

a chain, assuming only specification error in the original relations, will be a 33% standard error of the estimate. Thus, 68% of the predicted values will have an error of $\pm 33\%$ of the true value of Y.

Rather less dramatic but important error accumulations occur in all model systems. A typical early land use model generating inputs to a traffic model may have a marked chain structure which is deliberately intended to describe the sequences of decisions that are made during a phase of urban growth (Figure 3). However, since there is little evidence to suggest that decisions are normally sequential, it is preferable to design a system of simultaneous solution models.

An alternative approach is to introduce checks at each link in the model system network or flow chart. The ideal system would consist of several parallel models predicting distributions of the same activity, where each model would be a check on the other, similar in some ways to examining the alternative results from a Monte Carlo simulation. Alonso (1967) envisages a process of netting out weaker models by this sort of comparison. Where this is not feasible, checks can be made at each step using independent estimates of, for example, employment ratios, net residential density, and car ownership ratios. In the large urban simulation models such as

Pittsburgh (Steger, 1965), these checks prevent excessive error accumulation. Error in fact becomes additive rather than compounding.

d Evaluating linear models: Explicit measures of cumulative error rarely form part of model evaluation, but in most cases, the model is calibrated against observed patterns and judged for 'reasonableness' with the help of several statistical measures of goodness of fit. However the validity of the model will depend ultimately upon whether it suits its purpose, and then only within the context of the reasonableness and usefulness of its goals (Forrester, 1962, 115). But since this ultimate test may be rather elusive and imprecise, models are usually subjected to statistical testing.

This means looking at R^2 values (root mean square error values) and at the spatial pattern of residuals obtained from comparing observed and expected distributions. There are, however, several difficulties associated with this method. First statistical tests are not truly objective, and measures (such as RMSE values) are, as Forrester points out, quite arbitrary (Forrester, 1962, 122). Secondly there are no prescribed cut-off points or levels of statistical significance that will tell us whether or not a model is good enough to be used for prediction. This will depend entirely on the accuracy demanded by the problem, and it is often much easier to judge this if the results from a model are expressed in absolute terms.

However only rarely in the literature on urban models is great consideration given to the presentation of results with specific indications of what the confidence limits of the tests of significance are and what they mean in absolute terms. Hill (1965) reports that the EMPIRIC model had an R^2 of ·9. This represented an error of 3,000 resident population which was estimated as equal to 600 person trips in the rush hour, a volume that could easily be accommodated in one lane of a minor city street. Thus the important point is that most models are neither so accurate, nor evaluated in these understandable, and for most purposes, realistic terms.

A further complication is that R^2 and t-values may be misleading indicators of accuracy; this is because the stability (and also sensitivity) of the parameters is critical. If the parameters, which are predictors, are not constant over time, it is dangerous to use the model for prediction. As Lowry (1965) and Forrester (1962) emphasise, the longer the length of the forecasting period the more likely it is that the parameters will not remain stable. The reason for this is that the amount of noise in the model system multiplies through time. For example, shopping or journey to work habits may change their character over time, thus making long-term forecasts difficult. And this is particularly likely in areas of high growth or substantial change. It is probable that the parameters expressing relationships between variables in a stable industrial region in the United Kingdom will fluctuate less than those in an explosively developing city such as Caracas or São Paulo.

The sensitivity of parameters is also of interest to the model builder. The vulnerability of the model to small changes in parameter values will provide some clues about the nature of relationships in the model and the robustness of the model itself. If model results are highly variant under parameter changes, this may indicate that additional variables or restructuring of the model are necessary. If the model is reasonably invariant under these changes the model could be considered robust and suitable for forecasting or controlled experimentation. This kind of sensitivity analysis is commonplace in Monte Carlo simulation, but is rather infrequent in linear predictive models. One exception to this is Lowry's model for Pittsburgh which was examined for sensitivity and found to be 'not unduly sensitive to tinkering'.

e Conclusion on the linear model: One is forced to conclude that the linear model, despite its widespread use, is a highly restrictive model structure for urban investigation, even though many of the criticisms that have been made of it apply to other model structures. The overriding drawback of the structure is that it has been formulated for linear systems in the physical sciences and is rather unsuited to the non-linearities and complexities of urban growth. Yet because model structures for non-linear feedback systems are difficult to formulate (and the associated equations difficult to solve), the linear model provides an operationally straightforward structure which is readily available in computer program libraries. Nevertheless since the limiting assumptions of the linear model preclude elaborate simulation of urban growth, other methods (particularly probabilistic simulation) have been adopted which are more suitable for modelling complex processes. Generally these probabilistic models have not been designed for prediction and are thus less useful at present for forecasting the spatial pattern of urban development over the entire city.

2 Monte Carlo simulation models

a Discussion of the method: It is apparent that the urban system has a number of properties that do not make it amenable to simulation by deterministic equation systems. Time lags, trends, discontinuities and feedbacks, although as yet not clearly identified or measured, are characteristic of urban systems. Processes seem to be probabilistic and subject to possible random fluctuations and disturbances that cannot be incorporated in a linear model. Thus simulation, and particularly one branch known as Monte Carlo simulation, has frequently been proposed as the natural framework for both investigating and *experimenting with* urban phenomena (Garrison, 1962, 1966; Chapin, 1962, 1965; Harvey, 1966).

Simulation is described by Harling (1958, 307) as '. . . a technique of setting up a stochastic model and making experiments on the model.' Ackoff (1962, 348) describes simulation as a motion picture of a real world process which has four main uses.

1 Determining the values of controlled variables and testing the effects of changes in probability values and parameters.
2 Studying the transitional processes of a dynamic system. Orcutt (1961) points out that a simulation model can trace the system-wide effects of one input change in the model. The model can thus be used to trace step by step the actual flow of goods, information and transactions in a system and permits observation of the series of new decisions that take place.
3 Estimating the value of parameters with sensitivity analysis, by experimenting with the input values and parameters of the model.
4 Treating courses of action which cannot be formulated into the model. This is one of the primary functions of the method because it permits experimentation with complex phenomena. For example, in an urban model this might include investigating the effects of imposing certain zoning ordinances on a city, or estimating the traffic and land use changes that result from building a new road or shopping centre.

Monte Carlo simulation is a distinctive branch of simulation because it has the property of incorporating probability distributions. It has been widely applied in both the physical and engineering sciences and to a limited extent in the social and behavioural sciences, but has only recently been applied in urban geographical problems to which it would seem ideally suited (see Hammersley and Handscombe, 1964; Guetzkow, 1962; Cohen, K., 1960; Maas, 1962; Hoggatt and Balderston, 1963; Hägerstrand, 1965; Harvey, 1966, for discussion of the method and some applications).

b Monte Carlo simulation models and urban research: One of the earliest large urban simulation models was developed by Chapin, Weiss, Donnelly *et al.* at the University of North Carolina (UNC). Beginning with an investigation of the factors influencing the residential development of land (1962), they placed the land development process in a decision framework by considering the process as an orderly sequence of decisions. Donnelly, Chapin *et al.* (1964) have translated this concept into a probabilistic model in which secondary decisions—i.e. decisions to build on vacant land—became a function of four land development factors (the primary decisions): access to work, access to arterial roads, access to schools, and the availability of sewers. The urban area is divided into cells which are assigned attractiveness scores based on these four factors, and from these scores probability values and range of random numbers are derived. By drawing random numbers, households are assigned across the probability surface. After each iteration, however, the probability surface changes shape as the probabilities are recalculated according to the development that has taken place. For example, cells adjacent to newly developed cells or new highways obtain higher proabilities. This model has been calibrated with data from Greensboro, N.C. and further refinements are at present being built into the model, especially the simulation of decisions of land owners and developers (Chapin and Weiss, 1965).

Other workers, particularly in geography, have conceptualised the expansion of the urban area as a process of spatial diffusion following the work of Hägerstrand on modelling migration and the diffusion of innovations (1952, 1965). Morrill (1965) has designed a simulation model of the expansion of the urban fringe deriving his initial probability surface from observed patterns of growth in the study area near Seattle. He interprets this surface as a reflection of the accessibilities of land parcels to employment and services. Other probability functions describe the likelihood of development on different types of terrain, and the type of development (house, business and apartment) that is possible under different zoning conditions. Finally the density of development is probabilistically determined. The decision to develop a parcel of land is thus filtered through a number of probability functions until it is expressed as the location of a house, business establishment or apartment.

A rather different type of model was developed by Colenutt (1968) to simulate the spread of roadside land uses, particularly billboards, along a set of highways radiating out of a town. The probability surfaces for the road were derived from traffic flow data and were changed during the billboard diffusion process as the town expanded and land owners made decisions to adopt or resist billboards; rather in the same way as Morrill (1965) devised special landlord resistance rules for his Monte Carlo model of the diffusion of a ghetto. However the billboard model suffers from several of the limitations of these early urban simulation efforts (see also Pitts, 1964; Malm, Olsson *et al.*, 1966), and it is useful to look at some or the difficulties of building Monte Carlo models.

The first difficulty is the definition of the probability surfaces. The surfaces should theoretically be constructed or derived from some appropriate empirical observations (Harvey, 1966). If the actual surface of development is used to calculate probabilities, the model is merely simulating actual events, and with a sufficient number of simulation runs is bound to replicate precisely the actual development pattern. For this reason probability scores for the UNC model are calculated from the measures of cell attraction, and in the Colenutt model traffic flows provide the initial probability values. The probability surfaces therefore have some real meaning in terms of the process being modelled. But this reasoning does not seem to apply to Morrill's model of the urban fringe nor to other studies such as the Taaffe, Garner and Yeates model of the journey to work (1963) where probabilities are fitted by an iterative procedure to the actual journey to work pattern. This introduces a circularity into the model and seems to invalidate the use of the Monte Carlo method. Certainly it is difficult to see how the advantages of the method are exploited.

A second major difficulty recognised by all writers is to define the changes in the probability surface over time. How are these changes to be calculated? Again, resorting to use of time-series data on actual evolution of development patterns is circular, and hence some rules have to be set up. Often these are subjective rules based on observation and hunch, such as Morrill's terrain development function, or the decision rules adopted by Malm, Olsson and Wärneryd (1966) for building on rough terrain in their simulation of North Gothenburg. These rules, though, do not really solve the real problem which is to estimate a dynamic probability function that is based upon theoretical constructs or upon observed empirical regularities. Thus Donnelly (1962) suggests measuring changes in the attractions of different cells by using tax appraisers' assessment records. This problem becomes further confounded when probability functions merge and mix in the model so that the probability surface becomes extremely distorted and almost unrecognisable from the original surface. Consequently, in the later years of a residential simulation the initial probability surface itself may not play any important role in discriminating clearly between cells attractive to growth and those unlikely to be developed.

This leads to the third major problem—the management of the time dimension in the model. The numerous feedback and time-lag effects in urban growth make precise simulation of the *timing* of urban change, rather than the *spatial pattern* of development, very complicated. Generally either time-phased groups of housing units are fed in to the model in successive iterations as in the UNC model, or the numbers of housing units remain the same for each iteration and are later grouped into periods that correspond to real time. For example, in Hägerstrand's model of the diffusion of innovation (1967) the spread of information takes place at constant intervals when every carrier of information tells one other person. This is, as Hägerstrand recognises, rather unrealistic but he sees no way of avoiding this dilemma unless a dynamic diffusion function is derived. Hence most Monte Carlo simulations of spatial processes and urban growth are not truly time oriented but only simulate spatial patterns that result from processes that are implicitly assumed to operate over time.

c Testing and evaluating Monte Carlo models: The testing and evaluating of a Monte Carlo model are more difficult than testing a linear model, and are closely tied up with the purpose of the model. If the purpose of a particular residential model is to predict the population or amount of new construction in a set of subregions or cells, the desired output from the model is some definite numerical solution. Since a probabilistic model can produce a whole range of different solutions from successive runs of the model, choosing this value may be difficult. However one of the properties of the Monte Carlo method is that of convergence on an average solution after a number of simulation runs (Harvey, 1966), so that after perhaps 50 to 200 runs a numerical solution can be obtained.

If, however, the purpose of the model is to explore the nature of an urban process, such as the decision to develop vacant land, or to investigate the effects of land use controls on a growing suburban area, the average solution may not be particularly useful. What is more useful, is the examination of the range of results that a given set of probability functions generates. The distribution of results can be plotted for each cell in the urban area and measured for skewness (Colenutt, 1966). This kind of analysis can test the sensitivity of results over a series of simulations. It should be mentioned, however, that Harris (1962, 713) does not entirely agree with this approach and states that '. . . the interest in the results lies in the probabilistic distribution within a run of the model, and not primarily in variations between runs of the model.' In fact both types of variation are of interest to the model builder who is looking for both consistency of probabilities in the model and stability of the results.

Testing a Monte Carlo model consists, therefore, firstly of selecting a result or combination of results that are considered to be representative of the model, and secondly, of comparing these results with the actual pattern of city growth. This is not a problem peculiar to Monte Carlo simulation, but has assumed more importance in this type of simulation because there are fewer statistical tests of goodness of fit available for evaluating the model objectively. Various methods can be applied:

1 *Quadrat sampling*, or comparing the corresponding cells of the observed and expected distributions. Chi-square and Kolmogorov tests can be used for comparing patterns, but these do not test the comparability of the *patterns*, and the size and shape of the cells can lead to considerable deviations in the results. Harvey (1966) has shown that if various probability distributions are fitted to observed and expected patterns from simulation models it can be determined whether the distributions

produced by the model are random. For example, if a Poisson distribution can be fitted to the data then the distribution can be assumed to be random; but if a negative binomial or Neyman type A distribution can be fitted then the pattern is not random, and it is possible to infer whether the process operating is contagious or competitive over space. This may be important in a simulation where the first problem is to find out whether the model is producing a non-random pattern and the second to identify a statistical process that might be generating that pattern. But this kind of analysis still does not tell us anything about other properties of a spatial pattern—such as spatial trend, connectivity and contiguity.

2 *Contiguity and nearest-neighbour analysis* provide some measure of the nature of the connectiveness and grouping in a spatial pattern.
3 Comparing the *spatial autocorrelation functions* of observed and expected distributions will provide some idea of the extent of spatial trend and interdependence of spatial units.
4 *Spectral analysis* of the residuals of the series can be used to identify significant spatial frequencies that may have been omitted in the model formulation. Colenutt (1966) has attempted this for his linear diffusion model and found that the model had not taken into account at least one high energy spatial frequency.
5 *Subjective comparisons of the results.* Morrill (1965) in his ghetto model based his evaluation largely on inspection, and generally, other models are at least partially evaluated in this way.

It is apparent then that rigorous testing and evaluation of Monte Carlo simulation is laborious and expensive, and it suggests that building a simulation model is only worthwhile if the process being modelled definitely cannot be tackled by analytical methods. It is thus best suited to '. . . problems that involve a mass of practical complications' (Hammersley and Handscombe, 1964, 9), and then it should be quite clear what the precise purpose of the model is. It seems that Monte Carlo simulation is a research tool for exploring urban processes in depth rather than for simply making predictions. It is conceivable, however, that a Monte Carlo model might be used in conjunction with a series of econometric models for small area forecasting and plan evaluation. Curtis Harris (1966) suggests that his Markov model of residential growth might be used in this way.

VI Conclusion

Urban modelling is a relatively new field of interest to geographers, planners and other social scientists. It has largely arisen from transportation planning in the United States, but is now developing wider interests. More attention is being paid to modelling urban processes, such as the decision structure of residential development decisions, or the way in which persons on shopping trips evaluate their opportunities, or the factors that make up residential preferences. But these new modelling efforts are not necessarily producing either more accurate predictive models or socially valuable planning devices.

Better predictive models depend on three factors. Firstly, there is a need for more robust theorems to use in models, as Alonso (1967) and Levins (1966) suggest. A robust theorem is a statement about the relationships or interaction in the urban system which is reasonably invariant under different conditions—i.e. which is not unique to a particular city or time period. These theorems are derived from insights into urban interrelationships and from rigorous testing of hypotheses about urban growth and change. The second requirement is that the models themselves are robust. Robust models are those that are not unduly sensitive to small changes in parameter values. Rogers (1965) suggests that it is important to test models against different data sets removed in time and space from the calibration period and area. In this way more robust models can be derived from theorems that have some stability in time and space.

Finally, it would seem essential to record the predictive efficiency of a model over a forecast period and then return to the model and evaluate it in the light of this experience. This has been done with short-term econometric forecasting models of the United States economy (Suits, 1963), and it is now possible to do so with the first generation of traffic and land use models and even the early retail potential models. With this sort of feedback to the model builder, stronger models and more robust theorems can certainly be produced.

A major difficulty remains to be faced even if we develop models that do work well. None of the models so far discussed is able to evaluate the effect of locational decisions on the spatial pattern of real income in the city (see Harvey, 1970). Yet if our models cannot make these predictions, we are not only ignoring important urban processes but may find that models simply suggest the maintenance of trends that are increasing the inequalities in the distribution of land uses, and services, and the flow of goods. Such trends are undoubtedly undesirable and yet existing models completely ignore real income effects, social impacts, and service inequalities. This particular issue is perhaps the real frontier of urban model building.

VII Acknowledgements

I should like to thank David Harvey of Johns Hopkins University and Professor Haggett of the University of Bristol for their help during the preparation of this paper. Also I should like to thank Rodney White, Keith Bassett, Nathan Edelson and Marcia Merry of the University of Bristol, and Roger Downs of Pennsylvania State University for their suggestions and criticisms.

VIII References

Ackoff, R. L. 1962: *Scientific method: optimizing applied research decisions.* New York: Wiley, 364–404.

Alcaly, R. E. 1967: Aggregation and gravity models: some empirical evidence. *Journal of Regional Science* 7, 61–73.

Alonso, W. 1960: A theory of the urban land market. *Regional Science Association, Papers and Proceedings* 4, 149–57.

1967: Choosing and building models of prediction. *HRB Conference on Urban Models.* Hanover, N.H. (20 pp.)

Bartlett, M. S. 1947: The use of transformation. *Biometrics* 3, 39–52.

Berry, B. J. L. 1963: Commercial structure and commercial blight. *University of Chicago, Department of Geography, Research Paper* 85. (254 pp.)

1965: The retail component of the urban model. *Journal of the American Institute of Planners* 31, 150–5.

Berry, B. J. L., Barnum, H. G. and **Tennant, R. J.** 1962: Retail location and consumer behaviour. *Regional Science Association. Papers and Proceedings* 9, 65–106.

Blalock, H. M. Jr. 1961: *Causal inferences in nonexperimental research.* University of North Carolina Press.

Box, G. E. P. and **Tidwell, P. W.** 1962. The transformations of the independent variables. *Technometrics* 4, 531–50.

Box, G. E. P. and **Watson, G. S.** 1962: Robustness to non-normality of regression tests. *Biometrika* 49, 119–27.

Burgess, E. W. 1927. The determination of gradients in the growth of the city. *American Social Society, Publication* 21, 178–84.

Carrothers, G. A. P. 1956: An historical review of the gravity and potential concepts of human interaction. *Journal of the American Institute of Planners* 22, 94–102.

Cattell, R. B. 1966: Patterns of change: measurement in relation to state-dimension, trait change, lability, and process concepts. In Cattell, R. B., editor, *Handbook of multivariate experimental psychology*, Chicago: Rand McNally, 355–402.

Catton, W. Jr 1965: The concept of mass in the sociological version of gravitation. In Mansarik, F. and Ratoosch, P., editors, *Mathematical explorations in behavioural science*, Harwood, Illinois: Irwin, 287–321.

Chapin, F. S. 1967: Activity systems as a source of inputs for land use models. *HRB Conference on Urban Models*. (27 pp.)

Chapin, F. S. and **Hightower, H. C.** 1966: *Household activity systems: a pilot investigation*. University of North Carolina: Center for Urban and Regional Studies. (81 pp.)

Chapin, F. S. and **Weiss, S. F.** 1962: Factors influencing land development. *University of North Carolina, Institute for Research in Social Science, Urban Studies Research Monograph*. (101 pp.)

1965: Some input refinements for a residential model. *University of North Carolina, Institute for Research in Social Science, Urban Studies Research Monograph*. (68 pp.)

Cliff, A. and **Ord, A.** 1968: The problem of spatial autocorrelation. *Regional Science Association Meetings*, London.

Cochran, W. G. 1963: *Sampling techniques*. New York: Wiley. (413 pp.) (Second edition.)

Cohen, J. E. 1966: A model of simple competition. *Harvard University, Computation Center Annals* 41. (138 pp.)

Cohen, J. 1960: *Computer models of the shoe, leather and hide sequence*. Englewood Cliffs, N.J.: Prentice Hall (156 pp.)

Colenutt, R. J. 1966: *Linear diffusion in an urban setting*. Pennsylvania State University: unpublished Master's thesis.

1969: Linear diffusion in an urban setting: an example. *Geographical Analysis* 1, 106–14.

Cowan, P., Ireland, J. and **Fine, D.** 1967: Approaches to urban model-Building. *Regional Studies* 1, 163–72.

Cowie, S. 1969: *Residential preference functions and intraurban mobility*. University of Bristol: PhD dissertation. (In preparation.)

Digman, J. M. 1966: Interaction and non-linearity in multivariate experiments. In *Cattell, R. B.*, editor, *Handbook of multivariate experimental psychology*, Chicago: Rand McNally, 459–75.

Donnelly, T. G., Stuart, F., Thomas, G., Chapin, F. S. Jr and **Weiss, S. F.** 1964: A probalistic model of residential growth.

University of North Carolina, Institute for Research in Social Science, Urban Studies Monograph. (65 pp.)

Downs, R. 1969: *Consumer perception of shopping centres.* University of Bristol: PhD dissertation. (In preparation.)

Duke, R. D. and **Schmidt, A. H.** 1965: Operational gaming and simulation in urban research. *Michigan State University, Institute for Community Development. Annotated Bibliography* 14. (33 pp.)

Duncan, O. D., Cuzzort, R. P. and **Duncan, B.** 1961: *Statistical geography problems in analysing areal data.* Illinois: Free Press of Glencoe (191 pp.)

Ellis, B. 1966: *Basic concepts of measurement.* Cambridge University Press.

Ellis, R. B. 1966. *A behavioral residential location model.* Northwestern University: unpublished Master's thesis.

Ferber, R. and **Verdoorn, P. J.** 1962: *Research method in economics and business.* New York: Macmillan. (64 pp.)

Firey, W. I. 1960: *Man, mind and land: a theory of resource use.* Illinois: Free Press of Glencoe. (256 pp.)

Forrester, J. W. 1962: *Industrial dynamics.* Cambridge, Massachusetts: Press.

Garner, B. J. 1966: The internal structure of retail nucleations. *Northwestern University, Studies in Geography* 12.

Garrison, W. L. 1962: Toward simulation models of urban growth and development. In Norborg, K., editor, Proceedings of the IGU symposium on urban geography, *Lund Studies in Geography, Series B, Human Geography* 24, Lund: Gleerup, 91–108.

1966: Difficult decisions in land use model construction. *Highway Research Record* 126, 17–24.

Golledge, R. and **Brown, L. A.** 1967: Search, learning, and the market decision process. *Geografiska Annaler* 49B, 116–24.

Graybeal, R. S. 1967a: A model of retail development. *Western Section of the RSA meetings*, Las Vegas, Nevada.

1967b: A simulation model of residential development. *HRB Conference on Urban Development Models*, Hanover, N.H.

Guetzkow, H., editor, 1962: *Simulation in social science: readings.* Englewood Cliffs: Prentice Hall. (199 pp.)

Hägerstrand, T. 1952: The propagation of innovation waves. *Lund Studies in Geography, Series B, Human Geography* 4. Lund: Gleerup.

1965: Quantitative techniques for analysis of the spread of information and technology. In Anderson, C. A. and Bowman, M. J., editors, *Education and economic development*, Chicago: Aldine, 244–80.

1967: On the Monte Carlo simulation of diffusion. In Garrison, W. L. and Marble, D. F., editors, Quantitative Geography, Part I, *Northwestern University, Studies in Geography* 13, 1–32.

Haig, R. M. 1925: Towards an understanding of the metropolis: part II, the assignment of activities to areas in urban regions. *Quarterly Journal of Economics* 40, 402–34.

Hamburg, J. R. and **Sharkey, R. H.** 1961: Land use forecast. *Chicago Area Transport Study Report* 118. (143 pp.)

Hammersley, J. M. and **Handscombe, D. C.** 1964: *Monte Carlo methods*. London: Methuen.

Hansen, W. B. 1961: An Approach to the analysis of metropolitan residential extension, *Journal of Regional Science* 3, 37–56.

Hansen, W. G. 1959: How accessibility shapes land use. *Journal of the American Institute of Planners* 25, 73–6.

Harling, J. 1958: Simulation techniques in operations research. *Operations Research* 6, 307–19.

Harris, B. 1961: Some problems in the theory of intraurban location. *Operations Research* 9, 695–721.

1966a: *Notes on accessibility*. University of Pennsylvania: Institute for Environmental Studies. (12 pp.)

1966b: *Basic assumptions for a simulation of the urban residential and land market*. University of Pennsylvania: Institute for Environmental Studies. (35 pp.)

1966c: *Note on residential location in a subnucleated region*. University of Pennsylvania: Center for Environmental Studies. (3 pp.) (Mimeo.)

1967: Quantitative models of urban development: their role in metropolitan policy-making. *Conference on Urban Economics: Analytical and Policy Issues*, Washington D.C.: Committee on Urban Economics of Resources for the Future, Inc. (60 pp.)

Harris, C. C. 1966: A stochastic process of suburban development. *University of California, Berkeley: Center for Real Estate and Urban Economics, Technical Report* 1. (87 pp.)

Harris, C. D. and **Ullmann, E. L.** 1945: The nature of cities. *Annals of American Academy of Political and Social Sciences* 242, 7–17.

Harvey, D. 1967: Models of the evolution of spatial patterns. In Chorley, R. J. and Haggett, P., editors, *Models in geography*, London: Methuen, 549–608.

1968: Pattern, process and the scale problem in geographic research. *Transactions of the Institute of British Geographers* 45, 71–8.

1970: Social processes, spatial form, and the redistribution of real income in an urban system. *Paper prepared for the 22nd Colston Symposium, University of Bristol, on Regional Forecasting*, April 1970.

Herbert, J. D. and **Stevens, B.** 1960: A model of the distribution of residential activity in urban areas. *Journal of Regional Science* 2, 21–36.

Hill, D. M. 1965: A growth allocation model for the Boston Region. *Journal of the American Institute of Planners* 31, 111–20.

Hill, D. M., Brand, D. and **Hansen, W. B.** 1966: Prototype development of statistical land use prediction model for greater Boston Region. *Highway Research Record* 114, 51–70.

Hoggatt, A. C. and **Balderston, F. E.,** editors, 1963: *Symposium on simulation models: methodology and applications to the behavioral sciences.* Cincinnati: South-Western. (289 pp.)

Horst, P. 1966: An overview of the essentials of multivariate analysis methods. In Cattell, R. B., editor, *Handbook of multivariate experimental psychology*, Chicago: Rand McNally, 129–53.

Horton, F. E. 1967: The utility of trip forecasting models based on aggregate land use data. *Professional Geographer* 19, 319–22.

Hoyt, H. 1939: *The structure and growth of residential neighbourhoods.* Washington D.C. (131 pp.)

Huff, D. L. 1963: A probability analysis of shopping centre trading areas. *Land Economics* 53, 81–90.

1965: The use of gravity models in social research. In Massarik, F. and Ratoosh, P., editors, *Mathematical explorations in behavioral science,* Homewood, Illinois: Irwin, 317–21.

1966: A programmed solution for approximating an optimum retail location. *Land Economics* 42, 293–303.

Johnston, J. 1963: *Econometric methods.* New York: McGraw-Hill. (207 pp.)

Kain, J. F. 1962: *A multiple equation model of household location and trip making behaviour.* Rand Memorandum RM-3086-FF. (67 pp.)

Lakschmanan, R. T. 1964: An approach to the analysis of intraurban location. *Economic Geography* 40, 348–70.

Lakschmanan, T. R. and **Hansen, W. G.** 1965a: A retail market potential model. *Journal of American Institute of Planners* 31, 134–44.

1965b: A market potential model and its application to a regional planning problem. *Highway Research Record* 102, 19–42.

Lansing, J. and **Barth, N.** 1964: *Residential location and urban mobility: a multivariate analysis.* University of Michigan: Survey Research Center. (98 pp.)

Larson, H. J. and **Bancroft, T. A.** 1963: Sequential model building for prediction in regression analysis. *Annals of Mathematical Statistics* 34(2), 462–79.

Lathrop, G. T. and **Hamburg, J. R.** 1965: An opportunity-accessibility model for allocating regional growth. *Journal of the American Institute of Planners* 31, 95–103.

Leven, C. L. 1967: Trends in metropolitan growth and city form. *HRB Conference on Urban Development Models*, Hanover, N.H. (23 pp.)

Levins, R. 1966: The strategy of model building in population biology. *American Scientist* 18, 422–31.

Little, A. D. Inc. 1966: Model of San Francisco housing market. *San Francisco CRP, Technical Paper* 8, San Francisco: Little. (45 pp.)

Litton Industries, 1965: *Preliminary analysis for economic development plan for the Appalachian region.* Washington D.C.: Litton Industries. (197 pp.)

Lowry, I. S. 1964: *A model of metropolis.* Rand Memorandum RM-4035-RC. (135 pp.)

1965: A short course in model design. *Journal of the American Institute of Planners* 31, 158–66.

1967: Seven models of urban development: a structural comparison. *HRB Conference on Urban Development Models*, Hanover, N.H. (51 pp.)

Maass, A., Hufschmidt, M. M., Dorfman, R., Thomas, H. A., Marglin, S. A., and **Fair, G. M.** 1962: *Designing of water resource systems.* Cambridge, Massachusetts. E 33.91 D46.

Malm, R., Olsson, G. and **Wärneryd, O.** 1966: Approaches to simulations of urban growth. *Geografiska annaler* 48B, 9-22.

Matalas, N. C. and **Reiher, B. J.** 1967: Some comments on the use of factor analyses. *Water Resources Research* 3, 213–24.

McCarty, H. H. 1956: Use of certain statistical procedures in geographical analysis. *Annals of the Association of American Geographers* 46, 263.

Morrill, R. L. 1963: The distribution of migration distances. *Regional Science Association, Papers and Proceedings* 11, 75–83.

1965: Expansion of the urban fringe: a simulation experiment. *Regional Science Association, Papers and Proceedings* 15, 185–202.

Moser, C. A. 1958: *Survey methods in social investigation.* London: Heinemann.

Orcutt, G. H., Greenburger, H. M., Korbel, J. and **Ruitin, A.** 1961: *Micro-analysis of socio-economic systems: a Simulation study.* New York: Harper. (425 pp.)

Pearson, E. S. 1931: The analysis of variance in case of non-normal variation. *Biometrika* 33, 114–33.

Pitts, F. R. 1964: Scale and purpose in urban simulation models. *Conference on Strategy for Regional Growth*, Ames, Iowa. (12 pp.)

Putman, S. H. 1967: Intraurban industrial location model design and implementation. *Regional Science Association, Papers and Proceedings* 19, 199–214.

Ratcliff, R. U. 1949: *Urban land economics.* New York: McGraw-Hill.

Rhodes, T. and **Whitaker, R.** 1967: Forecasting shopping demand. *Town Planning Institute Journal* 53, 188–92.

Robinson, I. M., Wolfe, H. B. and **Barringer, R. L.** 1965: A simulation model for renewal programming. *Journal of the American Institute of Planners* 31, 126–34.

Rogers, A. 1965 : An investigation of retail land use forecasting models. *Bay Area Transportation Study*, Berkeley, California. (82 pp.)

Schlager, K. J. 1964 : Simulation models in urban and regional planning. *Regional Planning Commission, Wisconsin Technical Record* 3. (33 pp.)

Schmitt, R. C. 1954 : A method of projecting the population of census tracts. *Journal of the American Institute of Planners* 20, 102.

Schneider, M. 1959 : Gravity models and trip distribution theory. *Regional Science Association, Papers and Proceedings* 5, 51–6.

Scott, A. J. 1968 : *An equation system for the analysis of an urban complex.* University of Pennsylvania : Regional Science Research Institute. (Mimeo.)

Simmons, J. W. 1966 : Toronto's changing retail complex : a study in growth and blight. *University of Chicago, Department of Geography, Research Paper* 104.

Steger, W. A. 1964 : Analytical techniques to determine the needs and resources for urban renewal action. In *IBM Scientific Computing Symposium on Simulation Models and Gaming, Proceedings* 79–95.

1965 : The Pittsburgh urban renewal simulation model. *Journal of the American Institute of Planners* 31, 144–50.

Stewart, J. Q. 1950 : The development of social physics. *American Journal of Physics* 18, 239–53.

Stouffer, S. A. 1940 : Intervening opportunities : a theory relating mobility and distance. *American Sociological Review* 5, 845–67.

Stuart, A. 1962 : *Basic ideas of scientific sampling.* London : Griffin. (99 pp.)

Suits, D. B. 1963 : *The theory and application of econometric models.* Athens, Greece : Centre for Economic Research, Serbinis Press. (147 pp.)

Swerdloff, C. N. and **Stowers, J. R.** 1966 : Test of some first generation land use models. *Public Roads* 34, 101–9.

Taaffe, E. J., Garner, B. J. and **Yeates, M. H.** 1963 : *The peripheral journey to work: a geographical consideration.* Northwestern University : Transportation Centre. (125 pp.)

Tobler, W. R. 1966 : Spectral analysis of spatial series. *Fourth Annual Conference on Urban Planning and Information Systems and Programs*, Berkeley. (9 pp.)

Torgerson, W. S. 1965 : *Theory and methods of scaling.* New York : Wiley. (460 pp.)

Traffic Research Corporation, 1963 : *Reliability test report: POLIMETRIC model.* Boston Regional Planning Project. (44 pp.)

United States Department of Commerce, 1962 : *Traffic assignment manual.* Washington D.C. : Government Printing Office. (165 pp.)

1963 : *Calibrating and testing a gravity model for any size urban area.* Washington D.C. : Government Printing Office. (55 pp.)

Voorhees, A. M. 1961 : Development patterns in American cities.

Washington D.C., Highway Research Board Bulletin 293 (Urban Transport Planning), 1–8.

Voorhees, A. M. and **Associates,** 1966. *A model for allocating economic activities into sub-areas in a state.* Washington D.C. : Voorhees.

1967 : *Canberra land use transportation study: general plan concept.* McLean, Virginia : Voorhees, 91–7.

Voorhees, A. M. and **Shofer, J.** 1967 : Factors influencing work trip length. *Highway Research Record* 141, 24–38.

Wallis, J. R. 1965 : Multivariate statistical methods in hydrology : a comparison using data of known function relationships. *Water Resources Research* 1, 447–61.

Wendt, P. F. 1957 : Theory of urban land values. *Land Economics* 33, 228–40.

Wingo, L. Jr 1961 : An economic model of urban land for residential purposes. *Regional Science Association, Papers and Proceedings,* 191–205.

Wold, H. 1956 : Causal inference from observational data. *Journal of the Royal Statistical Society, Series A* 119, 28–50.

Zipf, G. K. 1947 : The hypothesis of the 'minimum equation' as unifying social principles : with attempted synthesis. *American Sociological Review* 6, 627–50.

Rethinking climatology

an introduction to the uses of weather satellite photographic data in climatological studies

---

by Eric C. Barrett

Contents

I The nature and purpose of climatology

THIS account aims to contribute positively to much of the field included in the peculiarly geographical science of climatology. Peculiarly geographical, that is to say, by traditional practice rather than by its theoretical contents and coverage, which overlap the boundaries between the atmospheric and earth sciences. Perhaps it is because most climatologists have claimed geographical rather than meteorological affiliations that climatology has been slow to benefit from the considerable advances made in meteorological observation and theory, particularly those since the end of World War II. Some of the greatest of these advances have resulted from the development of semi-permanent, Earth-orbiting satellite systems designed to increase and improve global observations of the atmosphere. The climatologist, although he is more interested than the meteorologist in generalities, often depends upon data acquired primarily for the other's use. It has been remarked recently (Vetlov, 1966) that only one tenth of the Earth's surface is adequately provided with observing stations even for present-day meteorological purposes. The inadequacy of the conventional observational network is, however, even more serious from the point of view of the climatologist for whom uniform and continuous data are not only optimal, but essential. The joint post-war developments in high-speed electronic computers and artificial Earth satellites have made possible between them the design and operation of remote sensing systems that are capable of making uniform measurements around the globe on a repetitive daily, or more frequent, basis.

The purpose of the present paper is to examine critically the potential contributions of some types of satellite data analysis to the solutions of problems of a climatological nature. It is the conviction of the author that satellite studies can play a large part in the rethinking or partial reorientation of climatology, whose classic writings in the past have been fundamentally descriptive and encyclopaedic in nature—for example Kendrew (1963) and A. A. Miller (1957)—and whose most widely publicised classifications have been based upon the results, rather than the qualities or mechanisms, of climates—for example the schemes devised by Köppen (see Wilcock, 1968) and Thornthwaite (1948). Table 1 attempts to summarise the spectrum of atmospheric sciences and their relationships with other disciplines. In the past climatology has been insufficiently related to atmospheric physics: it has been too preoccupied with the needs of non-meteorological consumers. The 'new climatologist' of the Space Age might do well to reassess his traditionally geographical patronage and to become more involved in physical, mathematical and

Table 1 The spectrum of atmospheric sciences

	1	2	3	4	5	6
Discipline	Aeronomy	Physical meteorology	Synoptic meteorology	Descriptive meteorology	Dynamic (physical) climatology	Descriptive climatology
Sciences	Chemistry Physics	Physics	Physics Maths.	Physics Maths.	Physics Maths. Geography	Geography Statistics
Subjects	Composition of atmosphere	Mechanics of atmosphere	Large-scale variations in atmospheric behaviour	Development of atmospheric systems	Climato-genesis	Climatic variations across space and time

even in chemical discourses. As in other physical geographical fields there has been a recent trend away from both qualitative description and measurement for measurement's sake, so in climatology more time and effort should be spent in developing understanding of *process*, here the cause of the differentiation between contrasting climates.

The atmosphere, moreover, is anything but a static medium; neither can 'climate' be considered as an invariable average state. Not until the theories of dynamic climatology have advanced significantly will the climatologist be able to contribute as usefully as he might to problems in operational meteorology as well as to those in geography. Sutton (1965) suggested that

> the short-lived deviations of the general circulation must be regarded as an intrinsic feature of atmospheric motion, much as the motion of any fluid tends towards turbulent fluctuations, whilst preserving its mean flow. The study of such variation . . . is the main theme of dynamic climatology.

This key aspect of climatology, involving and impinging upon the attempts of meteorologists to improve their forecasts and to extend them further into the future, is just the aspect that has been the most neglected. 'Mean flow' has attracted much attention in descriptive climatology; so, too, in meteorology, has the instantaneous state of the atmosphere; meanwhile short-lived circulatory deviations have attracted relatively little.

Perhaps the early growth of dynamic climatology was slow because data were inadequate and so fostered over-generalisation; if this was the case, then weather satellites must be regarded as highly important tools for climatological, as well as meteorological, uses, since they improve enormously the data available for processing by climatologists. Since satellite photographic studies have been much more numerous to date than infra-red analyses, the present account is almost exclusively oriented to photography. Ultimately, infra-red imagery must become the most valuable medium for atmospheric weather investigation, but that is yet to be.

II A survey of weather satellite systems

The development of weather satellite systems for the investigation of the atmosphere by remote sensing techniques was possible only when computer control of the complex spacecraft had become a practical proposition. In flight, as during launching, computer facilities are essential for the smooth operation of the very complex programmes of work at the ground Command and at Data-Acquisition stations (CDA for short). By September

1968, 21 American satellites designed specifically or importantly for sensing of the Lower Atmosphere had been successfully launched, only two launching failures marring the first 8½ years of satellite meteorology. The following notes summarise the more significant features of the first five weather satellite families.

1 TIROS (Television and Infra-red Observation Satellites)

Each TIROS satellite, having a simple hat-box configuration, measured about 107 cm in diameter and 57 cm in height. Power was supplied by over 9,000 solar cells arranged around the side walls. Most of these satellites, whose launch dates and orbital characteristics are summarised in Table 2, were equipped with television camera systems involving vidicon

Table 2 American weather satellites orbited prior to April 1968

Satellite		Launch date	Orbital period (mins)	Incln. to Equator (°)	Apogee (st. mi.)	Perigee (st. mi.)
TIROS	I	1 Apr. 1960	99·2	48·4	465·9	428·7
,,	II	23 Nov. 1960	98·3	48·5	453·0	387·0
,,	III	12 July 1961	100·4	47·8	506·4	461·0
,,	IV	8 Feb. 1962	100·4	48·3	524·8	441·2
,,	V	19 June 1962	100·5	58·1	603·9	366·4
,,	VI	18 Sept. 1962	98·7	58·3	442·1	425·3
,,	VII	19 June 1963	97·4	58·2	403·5	386·1
,,	VIII	12 Dec. 1963	99·4	58·5	468·0	435·0
,,	IX	22 Jan. 1965	119·2	96·4	1601·1	420·3
,,	X	2 July 1965	100·7	98·7	498·8	441·3
NIMBUS	I	28 Aug. 1964	103·5	98·7	578·0	263·0
,,	II	18 May 1966	108·5	98·1	860·1	680·2
,,	III	16 May 1968	Destroyed on launching			
ESSA	I	3 Feb. 1966	100·0	98·0	522·4	433·0
,,	II	28 Feb. 1966	113·4	98·0	882·5	840·9
,,	III	2 Oct. 1966	113·5	101·0	870·1	850·5
,,	IV	25 Jan. 1967	113·5	101·0	886·0	861·5
,,	V	20 Apr. 1967	113·6	101·6	922·0	861·0
ATS	I	7 Dec. 1966	24 (hrs)	2°N, 170°W*	22,236	22,224
,,	II	5 Apr. 1967	310	15·7	6,900·0	115·0
,,	III	5 Nov. 1967	24 (hrs)	0°, 47°W*	22,300	22,300
,,	IV	11 Aug. 1968	Experiment written off—(launching failure)		464·0	115·0

* Co-ordinates of geosynchronous position about the Earth's surface

camera tubes in the satellites, and television receivers at the CDA stations where the radio messages were rebuilt in pictorial form and were photographed by 35 mm cameras to preserve them. When the satellites were within radio reach of a CDA station, images could be transmitted directly to the ground. Around the greater part of each orbit, however, where CDA stations were far distant, in-satellite tape recording facilities accommodated sufficient image information in pretransmission form to provide data for most of the sunlit track next time a CDA station was overflown. Like all black and white weather satellite photographs, the picture data were transmitted from TIROS in unit numerals. Thus it was possible to achieve 10 different tones from black to white inclusive. Clearly this system placed limits upon the definition of the resulting images, but is still the most economic use of satellite and computer facilities. The latter are kept busy with an order of some 10^{12} data points each day from each operative weather satellite.

In TIROS VIII a new system of photo-reception was installed so that local users throughout the world might be able to obtain data directly from the satellites for use in forecast preparation, instead of the forecast correction that had been, for most national meteorological centres, the full extent of the usefulness of data received from the United States Weather Bureau several hours after the times of observations. This new Automatic Picture Transmission (APT) system is an important part of the present, fully operational satellite system. It involves the interception of automatically transmitted picture data from the satellites, and the rebuilding of these data on facsimile machines at ground complexes costing a tiny fraction of the third of a million dollars of the television CDA stations. The images from APT satellites acquire a dominant dot structure contrasting with the banded raster-line structure of television photographs. The resolutions of satellite pictures are functions of the camera lens angles (see Table 3), the satellite altitudes (see Table 2) and the read-out characteristics of the satellite/ground-receiver complexes. Table 4 summarises the best resolutions of photographic images of various kinds. Very high resolutions are technically possible, but for operational forecasting purposes a resolution of 2–3 km is apparently deemed adequate. For many research purposes it is not, and research workers should press for higher resolution pictures than those currently available.

The most important infra-red experiment in TIROS satellites measured outgoing radiation from the Earth and its atmosphere within the infra-red wave-bands specified in Table 5. Although the resulting data have not yet been evaluated fully, their very interpretation being difficult, they do suggest that similar recordings may eventually replace photographs as the most useful weather satellite data. The great potential advantage of infra-red data lies in their relative detail. The read-out and transmission of

Table 3 TIROS camera characteristics

Parameter	Narrow angle	Medium angle	Wide angle
Lens angle	12·5°	76°	104°
Lens speed	f/1·8	f/1·8	f/1·5
Shutter speed	0·0015s	0·0015s	0·0015s
Object area (viewed vertically from 400 miles)	70 miles sq.	450 miles sq.	750 miles sq.
Normal resolution	0·2 miles	1·5 miles	2·0 miles

Table 4 The best resolutions achieved by weather satellite data

TV photography	Satellite	Best resolution (st. miles)
VCS (narrow lens)	TIROS II	0·2 miles
VCS (medium lens)	TIROS V	1·0 miles
VCS (wide lens)	ESSA 3	1·5 miles
AVCS	NIMBUS I	0·5 miles
APT photography	ESSA 4	2·0 miles
Infra-red measurements:		
HRIR	NIMBUS	2 miles
MRIR	TIROS VII	30 miles
LRIR	TIROS IV	750 miles

satellite photographs is restricted by the use of the ten-unit digital code relating to black, white, and intermediate shades of grey, which of course restricts the tonal variations on the permanent photographic prints. By comparison, the infra-red data are accurate to about $+1$°K, across a range of some 100°K, from about 200–300°K.

2 ESSA (Environmental Sciences Services Administration) satellites

In February 1966 two satellites of the type pioneered by the 'cartwheeling' TIROS IX were put into orbit to inaugurate the first fully operational weather satellite system. As indicated in Figure 1, these cartwheel satellites roll around their orbital paths, and their cameras point towards the Earth through the side walls. Unlike the earlier TIROS photographs, all those obtained from ESSA are from preselected viewing angles, which

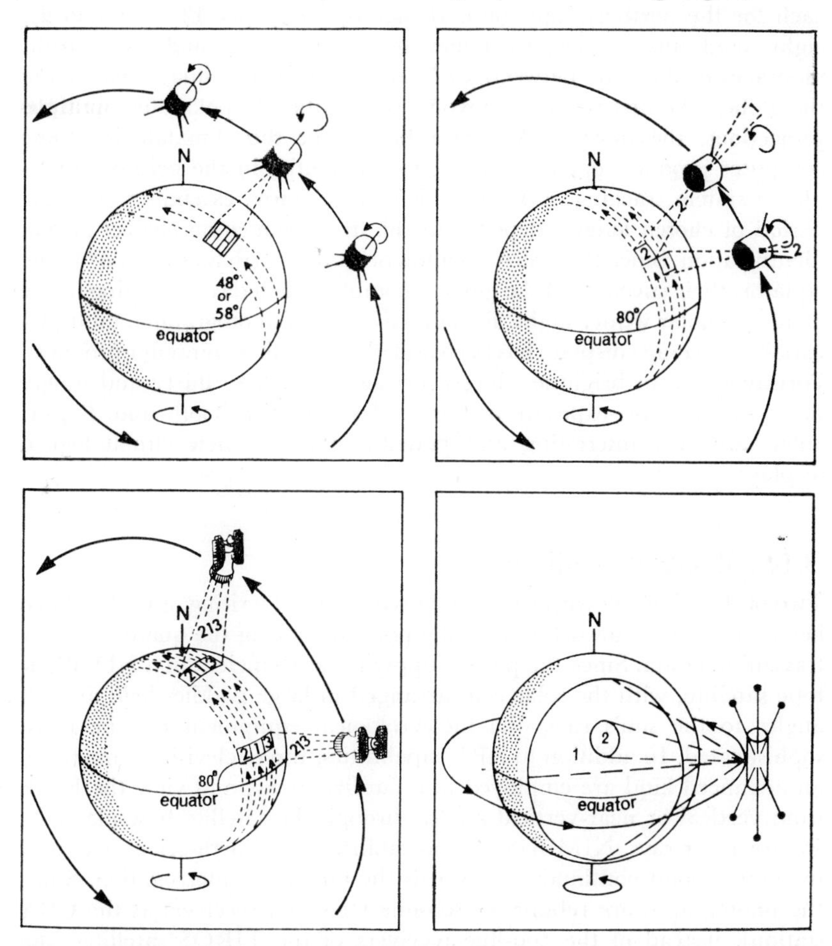

Fig. 1 Orbital characteristics of American weather satellites, namely TIROS (top left); NIMBUS (Bottom left); ESSA (top right), and ATS (bottom right).

remain constant around the entirety of each orbit. Since February 1966, two operational satellites, one television (for research purposes) and one APT (for local forecasting purposes), have been maintained in space by the launching of replacement satellites when their predecessors failed.

From the climatologists' standpoint, the greatest benefit contingent upon a predictable photographic angle is the range of computer-processed mosaics now produced on a daily basis by the United States Weather Bureau. These have been described more fully by Barrett (1968a); suffice it to say that the daily mosaics include two hemispheric montages (one each for the northern and southern hemispheres) (see Plates 1 and 2), eight quadrants of those montages at a larger scale, and six to eight montages of the zone between 30°N and 30°S of the equator encircling the globe. The first two types involve rectification to polar stereographic projections, the third to Mercator bases. For the climatologist, these computer products open up an exciting potential in the field of atmospheric system tracking, although this kind of study, extending over any period of chosen length, depends in the first instance on correct system identification. Since the ESSA satellites orbit the Earth at about 80° to the equator, their coverage of the globe is complete each day, excepting in the area(s) of polar winter night. Recently a range of photographic, multiple-exposure, average displays has been added to the other computer products. Portraying mean brightness levels for periods of five, thirty and ninety days, the multiple-exposure averages (described by Booth and Taylor, 1969) are very interesting and potentially useful, new climatological displays.

3 NIMBUS satellites

Two of these have flown successfully by the time of writing, and they have been especially noteworthy from the point of view of the quality of their transmitted data. Since the power supply is more efficient in a NIMBUS-type satellite, with the solar cells arranged in large paddles held at high angles to the sun's rays, the meteorological equipment can be more sophisticated. In addition to APT capabilities, three television cameras of an advanced kind are employed in a fan-like array, and view the Earth from vertical or near-vertical angles through the satellite base-plate. As Figure 1 portrays, NIMBUS is spin-stabilised and Earth-orientated, and its sensors point continuously towards the target for interrogation. Since the photographs are rebuilt on 800-line television receivers at the CDA stations, instead of the 500-line receivers of the TIROS satellites, the photographs are of a relatively high quality, and have been used widely as the base data for many detailed meteorological and climatological studies.

In the infra-red field, NIMBUS satellites have provided very useful and interesting data through a high resolution infra-red radiometer (HRIR) recording radiation between 3·5–4·2 μ, a sharply defined atmospheric water vapour 'window' wave-band. The resolutions of these data have compared very favourably with those of most weather satellite photographs.

4 COSMOS satellites

Although it is difficult to obtain up-to-date information from the Russians regarding their weather satellite programme it is known that some members of the very large and cosmopolitan family of COSMOS satellites have been meteorological satellites rather similar to the NIMBUS type. For example, COSMOS 122 was launched on 15 June, 1966, into a circular orbit at an altitude of *c* 650 km, and at 65° to the Equator. Two cameras were employed to photograph slightly overlapping strips on either side of the sub-satellite track, covering between them a strip some 1,000 km wide. A particularly good photo-coverage of the U.S.S.R. was achieved thereby.

The infra-red equipment included an HRIR sensor akin to those of NIMBUS I and II, and a three-channel MRIR recording in the wave-bands listed in Table 5.

Table 5 Medium resolution infra-red wave-bands investigated by weather satellites

Channel	Nominal spectral range (microns)	Measurement
TIROS		
1	6·0–6·5	Absorption by water vapour
1 (TIROS VII)	14·8–15·5	Absorption by carbon dioxide
2	8·0–12·0	Major 'atmospheric window'
3	0·2–6·2	Reflected solar radiation
4	7·0–30·0	Long-wave radiation back to space
5	0·55–0·75	Reflected visible solar radiation
NIMBUS		
1	6·4–6·9	Absorption by water vapour
2	10·0–11·0	'Atmospheric window' readings
3	14·0–16·0	CO_2-absorption
4	5·0–30·0	Emitted long-wave radiation back to space
5	0·2–4·0	The intensity of reflected solar radiation
COSMOS		
1	0·3–3·0	Reflected solar radiation
2	3·0–30·0	Long-wave radiation back to space
3	8·0–12·0	Major 'atmospheric window'

Later Russian experiments have involved photographing cloud patterns through red and green filters on successive orbits. By overlaying transparencies of the photographs, the changes in cloud patterns become clearly apparent in the distributions of reds, greens and browns.

5 ATS (Applications Technology Satellites)

These versatile, multi-purpose satellites represent a second generation in that they are designed and built to act as weather communication satellites, linking various ground stations, as well as meteorological observatories. They differ radically from earlier types in that they are geosynchronous or geostationary, apparently fixed at 22,300 miles above preselected points on the Earth's surface.

Their wide-angled cameras photograph the entire visible disc of the Earth (approximately one third of its surface) in each frame, and the same areas are photographed repetitively throughout the day. Clearly there is great scope for the study of diurnal dynamic cycles in the atmosphere in so far as these are made visible and are traced by cloudiness. The ultimate in time-lapse cloud films has been prepared by Suomi and his co-workers at the University of Wisconsin, employing ATS wide angle camera products for the macro-scale view, and narrower angle camera products to focus attention in more detail upon smaller regions. Again it is not difficult to conceive purely climatological studies, stemming from considerations of cloud types and cloud cover, that could be based on films of this kind.

The other ATS innovation that promises to aid the student of cloud cover is known as the multicolour spin-scan camera system. This, in ATS-3, permitted for the first time the retrieval of colour television pictures from weather satellites, and the results have been startlingly good.

6 Retrospect and prospect

This section has outlined concisely the development of weather satellite observational capabilities up to mid-1968. It affords background information essential for the studies described later, since the ultimate limitations of these studies are largely dictated by the design characteristics of the satellites and their meteorological equipment. For further details on these technical matters the reader could consult profitably the following references. General accounts of spacecraft systems include Widger (1966) and Barrett (1967a), both of which deal with the period from the launching of the first weather satellite in 1960 to the inauguration of the operational ESSA system early in 1966. More recent systems and satellite/computer products are explained by Barrett (1967b and 1968a), and future systems by Stein (1967). The NIMBUS I and II Users' Guides (NASA, 1965 and 1967) contain much valuable information relating to these advanced satellites, and to the ranges of data available from them. The article by Lehr (1962) summarises methods of data archiving and retrieving, and those methods largely apply still.

It may be concluded that in the earlier years when satellites were space-orientated, the *variability and obliquity* of their viewing angles of the Earth constituted the major obstacles to worthwhile interpretation. Since then, with cartwheel-type and Earth-orientated satellites, the major limiting factor has been the *resolution* of the data. Whilst a resolution of 2–3 km may be satisfactory from the point of view of a synoptic meteorologist, and even from that of some climatologists whose interests centre upon broad-scale problems, it will be demonstrated later that fine detail is highly desirable for most climatological studies. It is to be hoped that research workers may be able to prompt the development and operation of more acute sensing devices.

Two further developments that would seem to be highly desirable concern night cloud photography, and stereoscopy. The ill-fated ATS-3 satellite carried a low-light camera designed to photograph the Earth and its atmosphere on the night side of the Earth. Future satellites of this type will be similarly equipped, and these will facilitate the study of diurnal cloud cycles. Meanwhile the only published article on stereoscopic interpretation of weather satellite photographs arose from the chance optimisation on NIMBUS II APT pictures of resolution, tonal contrast, the base/height ratio, and the exposure interval, thus permitting three-dimensional examination (Ondrejka and Conover, 1966). This study afforded a useful corroboration of earlier works on cloud recognition and interpretation, amongst which those outlined in the next section figured prominently. There would appear to be no specific plans to include stereo-cameras as such in weather satellites in the near future but, according to Oliver (personal communication, in 1968), favourable decisions in this direction might be taken soon.

III The interpretation of satellite cloud photographs

1 The recognition and identification of clouds classified in terms of their appearances

Prior to the advent of weather satellites different cloud types were identified usually in terms of the forms and altitudes of individual clouds or cloudlets. (See, for example, the systematic description of cloud types in Berry, Bollay and Beers, 1945). Since the individual components of many types of cloud fields are, however, too small for current satellite photographs to resolve, new schemes of cloud classification had to be developed in the early days of satellite meteorology which would take into account the appearances of the different types of cloud fields themselves. On most TIROS and ESSA satellite photographs it is possible to recognise and to identify individual cumulo-congestus and cumulo-nimbus clouds (Ericksson, 1964) and on many NIMBUS photographs streets of cumulus are visible also, but, not surprisingly, even these adopt shapes that cannot be appreciated easily from ground or aircraft observations.

The earliest attempts to classify generically the cloud types portrayed by weather satellite photographs were published by Conover (1962a, 1963), and these have become well-proven, even classic, schemes. Conover suggested that it was possible to identify a wide variety of cloud types, under moderate to good photographic conditions, by employing six descriptive criteria, which may be summarised and explained briefly as follows. For a more detailed discussion of these points, see Barrett (1967b).

1 *Cloud brightness*. This has been shown to be affected by a number of meteorological factors, such as the depth and composition of the clouds (Barrett, 1967b), the angle of the sun's illumination of the cloud fields, and various technical aspects of the satellite ground system hardware complexes (Jones and Mace, 1963).

2 *Cloud texture*. Conover distinguished several textural appearances—smooth or fibrous (cirriform clouds), smoothly opaque (low stratus and fog), mottled or irregular (cumuliform or strato-cumuliform), smooth but ragged with inter-digitating brightness bands (stratiform clouds), and various hybrid forms.

3 *Vertical structure*. This could be deduced sometimes from oblique photographs of the Earth-orientated TIROS variety, by the occurrence of overlapping cloud layers of different textures. This is less feasible with the more nearly vertical photographs of recent satellites, but Conover and his co-workers, as noted above, made the important discovery that, more by chance than design, some of the NIMBUS II photographs could be examined stereoscopically if processed suitably. This has aided the growth of expertise in the interpretation of 'flat' photographs.

   Also important in the context of vertical structure interpretation have been the studies of cloud shadows. Shadow sometimes helps the photo-interpreter to distinguish the taller clouds under conditions of oblique illumination (World Meterological Office, 1966), and to identify the presence and alignment of high cloud, especially of the jet-stream cirrus type, by virtue of the patterns of their shadows cast on lower clouds or the surface of the Earth (see, for example, Whitney, 1966)

4 *The form of cloud elements*. These, the smaller distinguishable units on satellite photographs, are helpful for type recognition in areas of

cumuliform and strato-cumuliform rather than cirriform or stratiform cloud fields. It has been claimed that the forms of those elements are closely related to the directions and strengths of the thermal winds through the depths of the atmosphere in which they are developed (see Hubert, 1963).

5 *The patterns of cloud elements.* These are known to vary in relation to a wide variety of atmospheric and topographical factors, but there is greater uncertainty of interpretation at this more detailed level of study than at any other: the descriptive classification of these features is incomplete. Atmospheric and other factors involved in the differentiation of cloud elements include the strength of air motion, vertical and horizontal wind shears, the organisation of stability and instability, and the roughness or smoothness of the underlying Earth surface.

6 *The sizes of the elements and/or patterns.* Since it is an empirical fact that some types of cloud organisation occur across a very wide range of scales, the sizes of clouds and cloud systems must be taken into account in any generic interpretation.

Figure 2 illustrates the relationships between the various genera and species of cloud, identified by Conover (1963). It can be seen that most types of clouds identifiable from the Earth's surface are also distinguishable from space, but it must be remembered that in areas of layered clouds it is often not possible to differentiate the superimposed or overlapping layers. It is possible that integrated aircraft and satellite observational studies might permit the recognition of a wider variety of stratiform cloud species than in the Conover classification, but generally speaking his scheme is a good one, and should be altered by addition only.

2 The classification of clouds in terms of the modes of their development

Once the recognition of morphologically different cloud types had been seen to be possible, the next problem of interpretation involved a greater emphasis on the satellite-recorded clouds. Cloud patterns and cloud elements are generally related to a variety of atmospheric processes and stimuli (Barrett, 1968a), which include stability and instability, the distributions of water vapour and solar energy, the strengths and relative strengths of vertical and horizontal motions in the troposphere, and the wind shears in both planes. Cloudiness can be regarded as a useful tracer within the constantly changing fluid medium of the atmosphere, and its shapes and forms arise in ways that both indicate and integrate the patterns of distribution of other atmospheric elements.

The first attempt to develop 'an ultimate standard (genetic) classi-

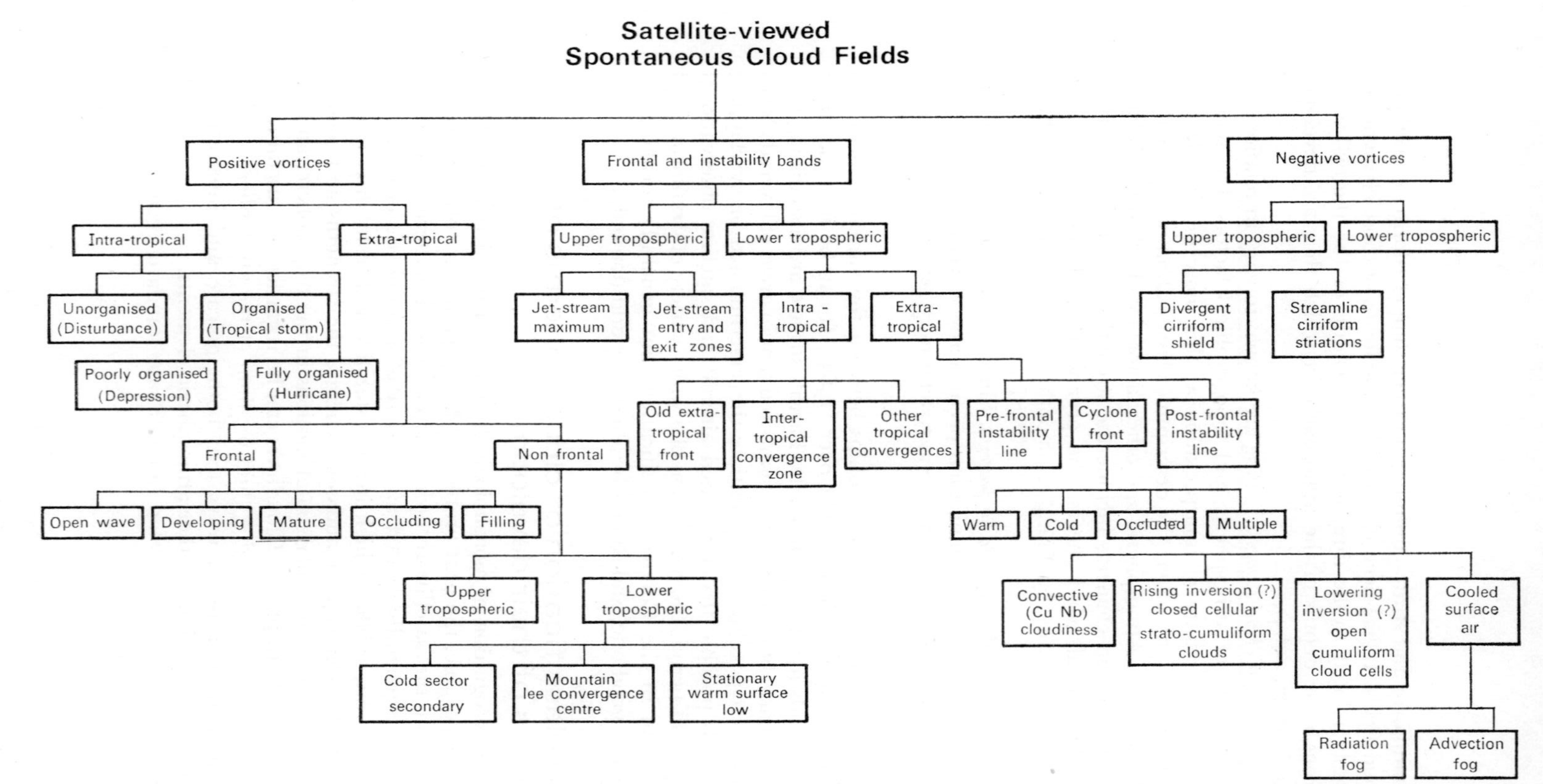

Fig. 3 A basis for a generic classification of clouds portrayed on satellite photographs. Section 1 classifies cloudiness organised dominantly by processes in an atmosphere undisturbed by Earth topography.

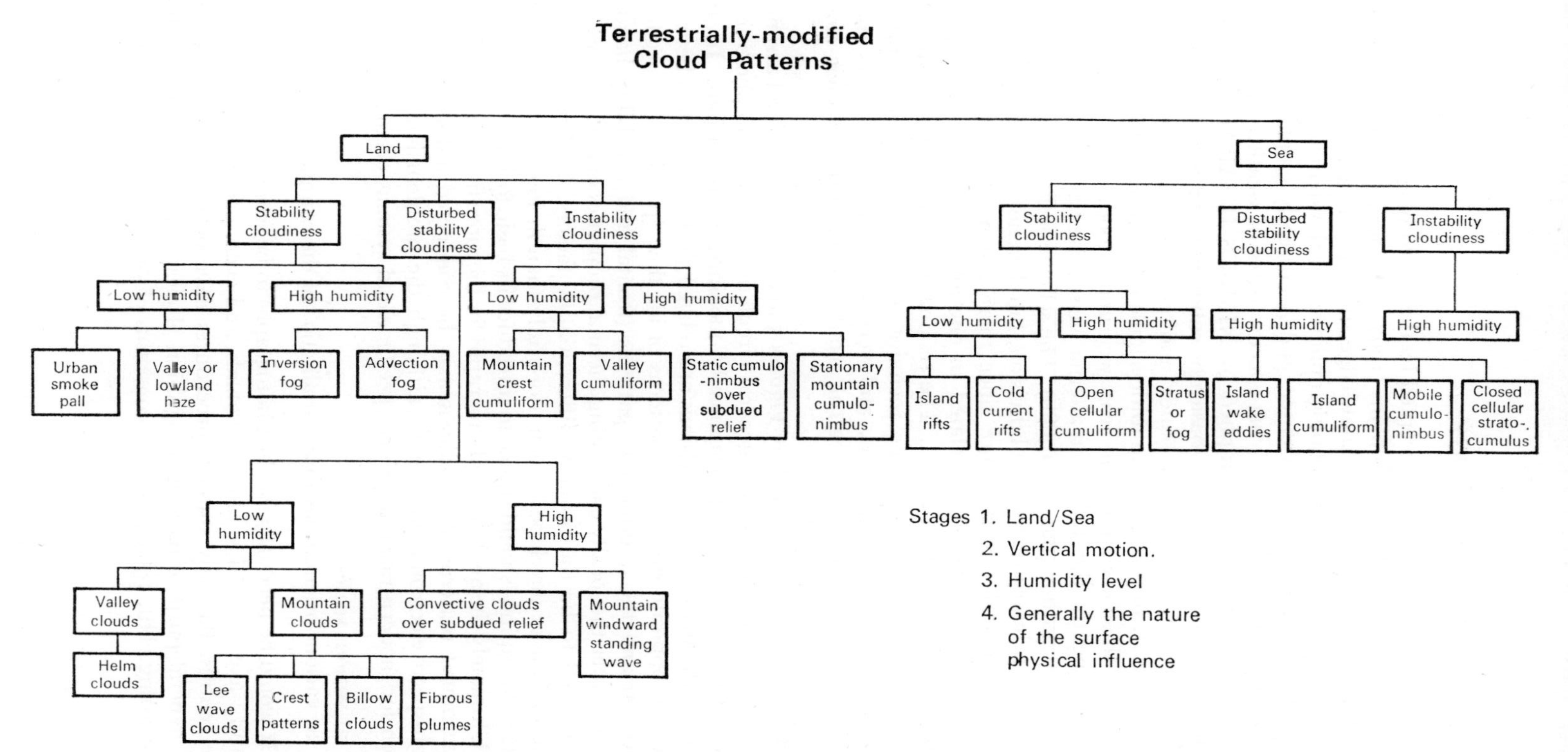

Fig. 4 Section 2 of a generic classification of clouds portrayed on satellite photographs. This suggests a subdivision of clouds and cloud rifts stimulated largely by variations in terrestrial surface.

fication that will be widely accepted by the meteorological community as a whole' was published by Hopkins (1967). This was designed not only as a contribution of academic knowledge, but also as a framework for a computer-operated satellite photographic retrieval system at the American National Weather Records Center, and is currently employed to assist researchers interested in studying specific types of weather systems and cloud arrangements photographed by NIMBUS II (ARACON, 1967). Hopkins' scheme, however—basically subdivided in terms of 'vortical cloud features', 'major cloud bands' and 'general cloud features' —does not do credit, in particular, to anticyclonic cloud fields, which is a great pity since these are among the least well understood.

Consequently, the basis of an alternative scheme has been suggested by the present author (Barrett, 1968b). The scheme seeks to be comprehensive in its primary stages of subdivision, but does not claim to be so lower down its logical tree structure since there are many arrangements of clouds that have not been related to one another and to atmospheric process sufficiently well as yet. The residual cloudiness that escapes the net of the classification filter includes unusual and/or incompletely explained cloud arrangements, and very useful research studies, both meteorological and climatological, could concentrate on their closer examination.

The scheme, as illustrated by Figures 3 and 4, separates cloud arrangements that owe more to the general circulation of the atmosphere from those that represent topographically modified perturbations. In the former, the first-order subdivisions are related to the differences between the two opposite spiraliform patterns of air flow (positive and negative vortices) and the intermediate patterns (frontal and instability bands). In the latter the locally contrasting thermal properties of land and sea are considered to be the primary differentiating factors. Clearly the classification can be developed much more fully than in Figures 3 and 4, but these illustrations should suffice to introduce its basic form.

Ultimately the analyses of photographs for the more frequent genetically determined patterns of cloudiness may be conducted automatically by high-speed electronic computers, although the problem is a severe one even for modern machines since specific cloud systems occur:—

1 at various scales
2 at various orientations through all 360° of the compass.

It would be necessary for the photographs to be scanned parallel to the *x* and *y* axes, as well as along diagonals, and even, perhaps, spirally from the centres. Basically suitable procedures have been developed already for general purpose pattern recognition (Rosen, 1967), and for weather forecasting (Hu, 1963), but the consensus of present opinion is that the likelihood of developing operational computers especially for these complex tasks is further distant than earlier anticipated. For the immediate future, human recognition and identification of clouds and cloud organisations remains the key to the use of satellite photographs in meteorology and climatology.

3 Satellites nephanalyses

In earlier, conventional meteorology, 'nephanalysis' involved the study of cloud types, patterns and amounts seen from the ground. Since the

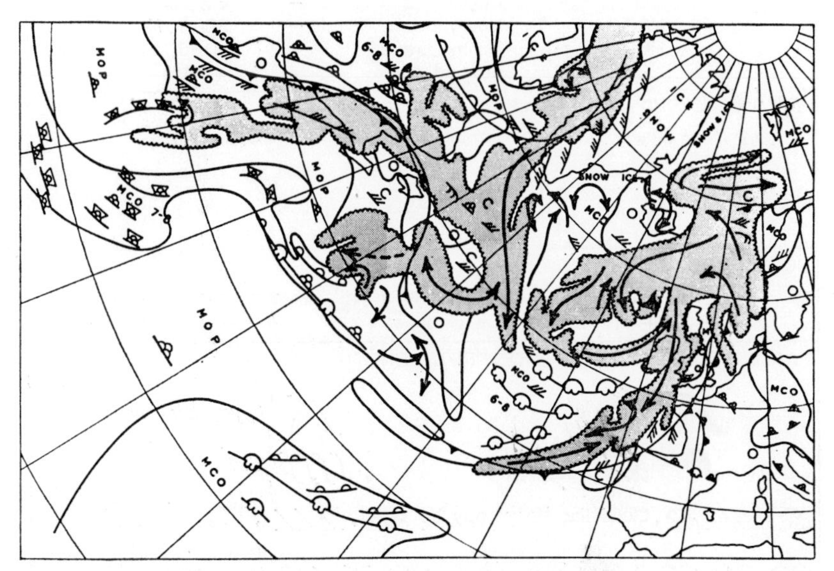

Fig. 5. A sample nephanalysis summarising cloudiness across the North Atlantic on 17th July, 1967.

first days of satellite meteorology, however, that term has acquired a more specialised connotation with the routine preparation of simplified cloud charts from satellite photographs. These nephanalyses, of which Figure 5 is an example, are prepared primarily for forecasting uses and present the rich photographic information in simplified form by means of a newly developed code of symbols, illustrated by Figure 6.

Nephanalyses were designed originally for near real-time transmission of salient cloud information to synoptic weather forecasting offices throughout the world via the international weather radio facsimile network, and they are still useful today in two major respects:

1 They provide national forecasters throughout the world with independent checks on their own interpretations of APT pictures.

2 They are useful research documents (see section V), partly because of their generalised nature, and partly because the various categories of cloudiness depicted on them are expressed quantitatively in terms of percentages of overcast.

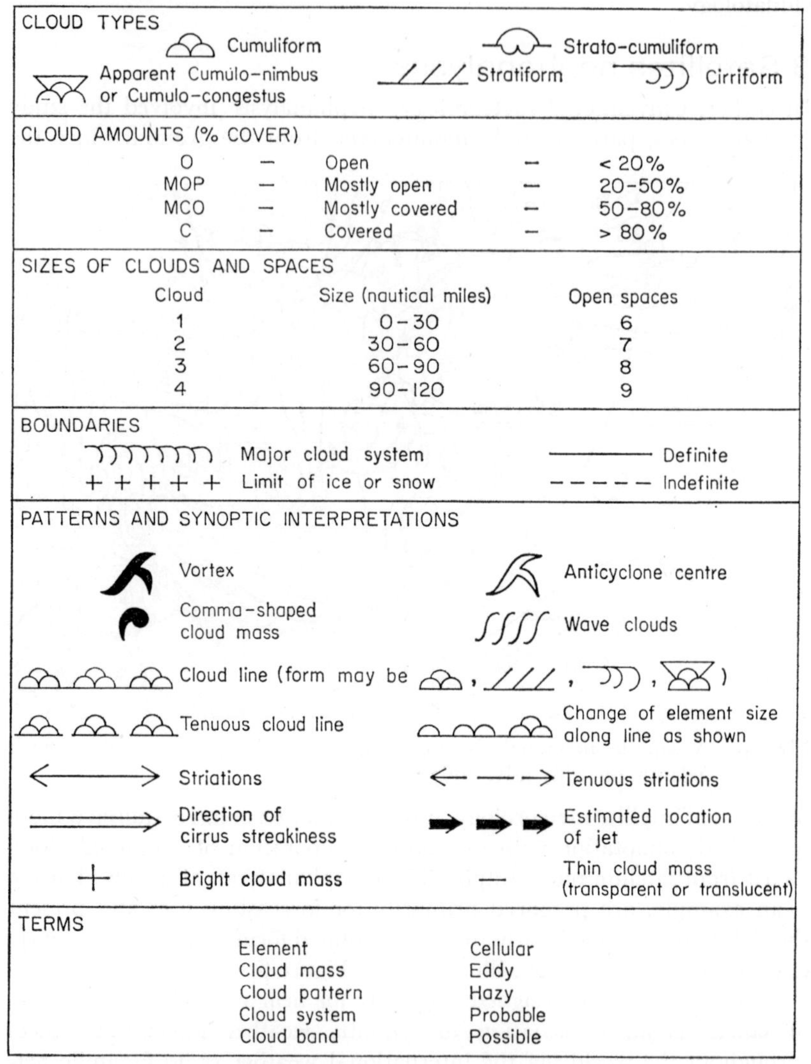

Fig. 6 The code of conventional nephanalysis symbols. *From Barrett* (1967b)

Admittedly the nephanalyses are still prepared by eye, but the photo-analysts involved in the task are highly experienced and the charts are quite reliable. As Figure 6 shows, they portray cloud types classified generically (for example, stratiform, cumuliform, strato-cumuliform, cumulonimbus and cirriform clouds), genetically interpreted features (for example jet-streams, vortices, etc.), and give quantitative assessments of strengths of cloud cover.

One of the problems attending the research use of satellite photographs has been the embarrassing complexity of the pattern they portray. It has been estimated that nephanalysis compilation results in a reduction of the original data bulk of the photographs by about two orders of magnitude; thus a considerable amount of detail is lost, but to the climatologist interested in the macro-scale this fact can be a source of considerable relief.

IV Satellite studies of weather systems

Dynamic climatology is much more concerned with the structures and resulting effects of individual weather systems than is descriptive climatology, because more emphasis is placed in the former upon the mechanisms of climate, climatic differentiation from place to place, and climatic variation across time. Many satellite studies have examined organised systems of cloudiness and weather, and this section summarises briefly some of the more important of them.

Since the first weather satellites were intended to keep watch on severe tropical low pressure vortices, and since these small, brightly reflective systems are amongst the best defined on satellite photographs, it is not surprising that some of the earliest weather system studies focused upon them, and that knowledge of them has been significantly advanced. Amongst these studies, that of Fett (1964) is especially worthy of note. This illustrated a mode of development of severe, well-organised vortices from easterly wave flow patterns (see Figure 7), and incorporated the newly confirmed outer convective band of cloudiness, the annular subsidence zone, and trailing convective cloudiness into a mature hurricane model. The relationships between these features and the other longer recognised features of a hurricane are depicted by Figure 8. A similar model is discussed at greater length by Barrett (1967b), although Figure 8 includes additionally an indication of the so-called primary energy cell (EC), whose existence in the forward right-hand quadrant of many maturing hurricanes has been confirmed by both visual and radar observations. This energy cell appears to be the primary link between the lower inflow and the upper outflow of the storm (United States Weather Bureau, 1965). Studies of individual storms led Malkus *et al.* (1960) to suggest that the cumulo-nimbus towers involved in this link covered only a small percentage of the radii, perhaps as low as 1%. These are very

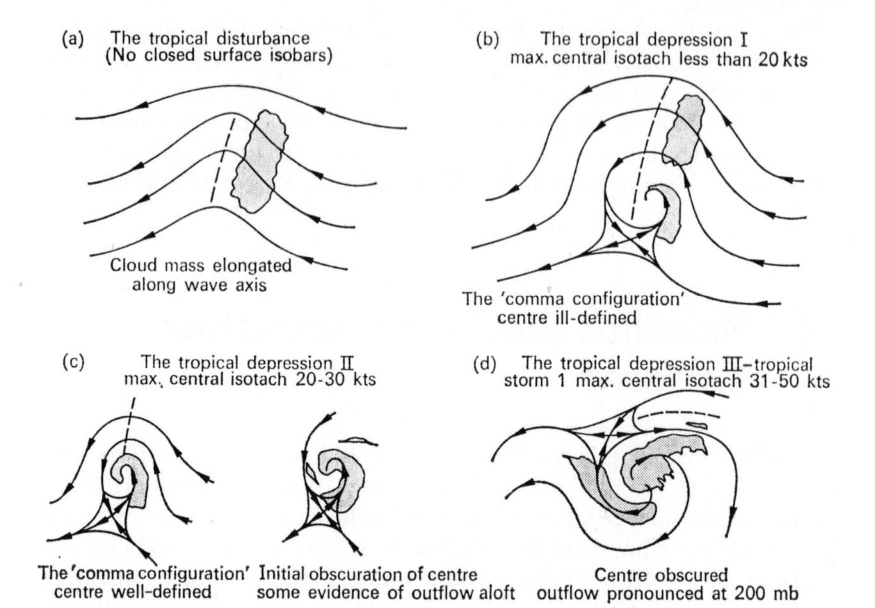

Fig. 7 Early stages in the development of a hurricane from an easterly wave streamline pattern. *From Fett* (1964)

difficult to identify on satellite photographs without the evidence of the radar screen.

The potential benefits of time-integrated radar and satellite photographic studies of various types of weather systems can scarcely be overestimated. Satellites are able to provide graphical representations of cloud cover, and radar is able to indicate the patterns of precipitation falling from the areas of clouds. Indeed it is true to say that, until weather radar revealed the spiral nature of hurricane-rainbands, no one had been led, either by theoretical considerations or the examination of satellite photographs or rainfall records, to suspect that the rain-bands might be spiral in plan (B. I. Miller, 1967). Whilst it is often possible to identify precipitation bands within extra-tropical depression fronts on satellite photographs, it is much more difficult to do so within the overcast discs of hurricanes.

Another co-operative programme of research which is capable of development in the future was initiated by Timchalk (1965). During the

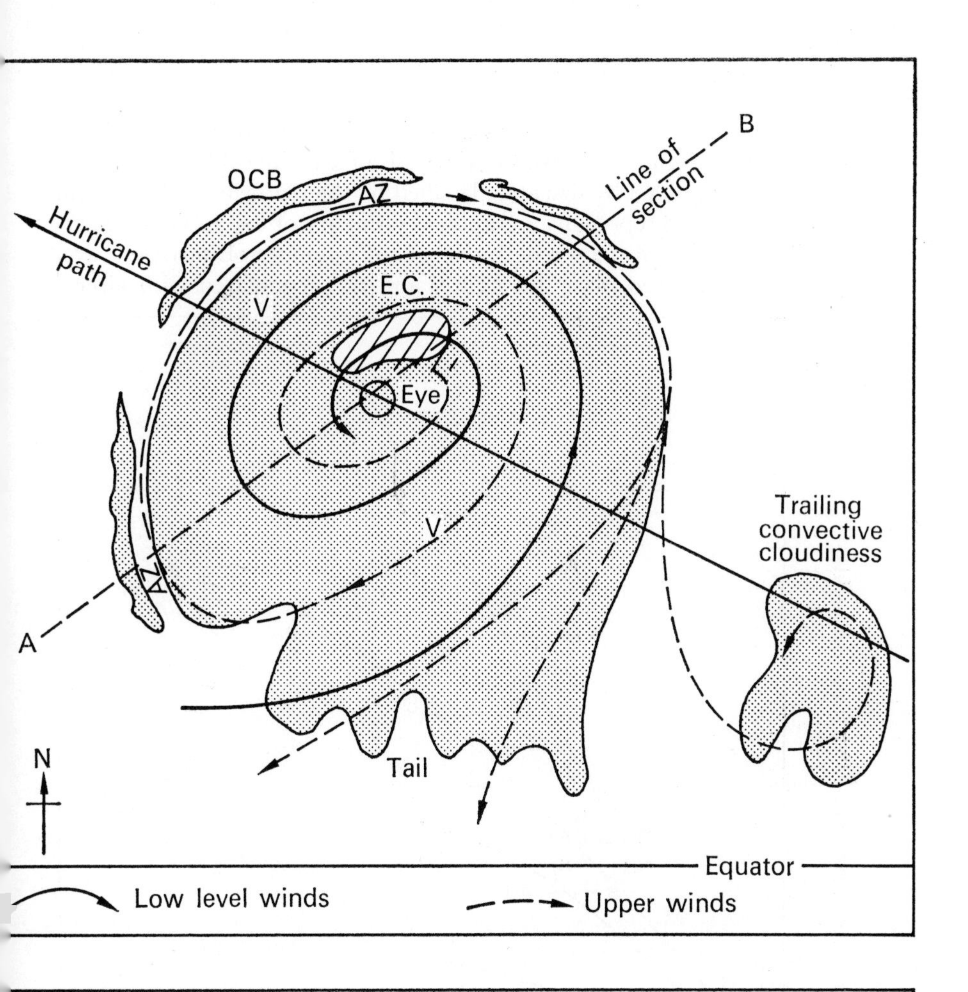

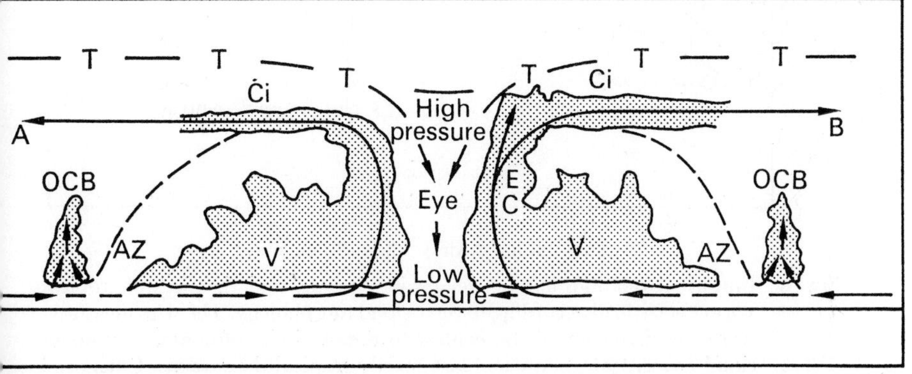

Fig. 8 A model of a mature northern hemisphere hurricane incorporating features discovered and confirmed by satellite studies. *T*: tropopause; *Ci*: cirrus; *V*: hurricane Vortex; *OCB*: outer convective band; *AZ*: annular zone; *EC*: primary energy cell. *From Barrett* (1967*b*)

hurricane season of 1962, wind speeds in a large number of hurricanes were measured and recorded. On the basis of satellite evidence the storm systems were graded by size and classified by the maturity of their organisations. Figure 9 represents a graphical summary of the findings. During the following hurricane season independent estimates of hurricane wind speeds were derived from this figure, and then compared with aircraft

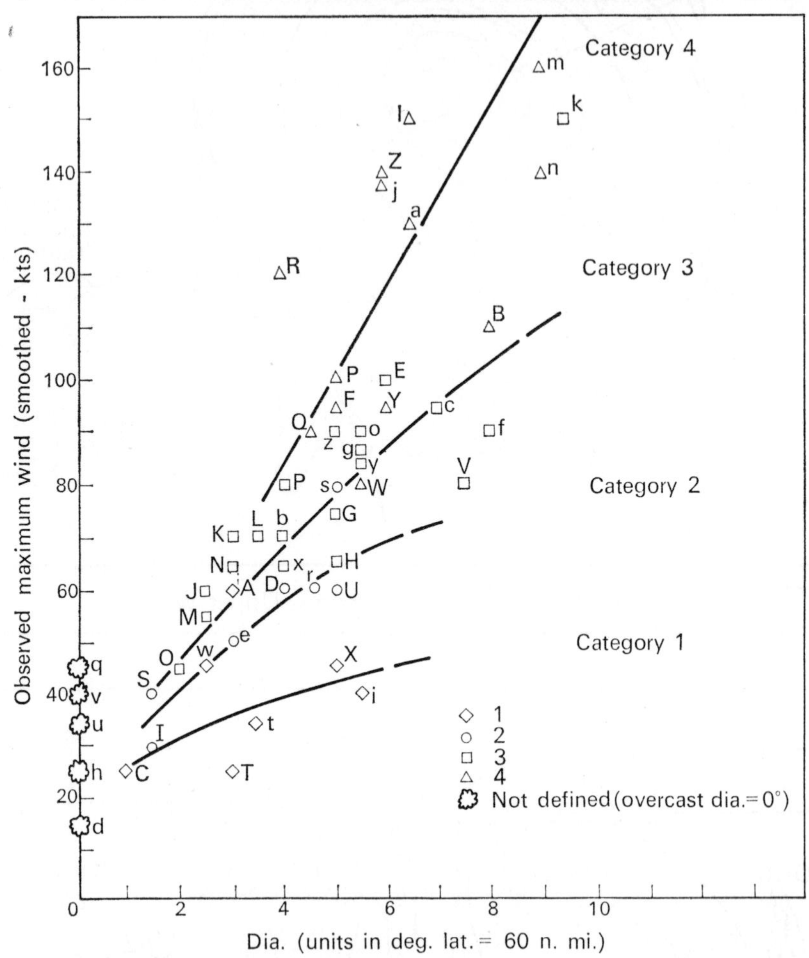

Fig. 9 Empirical relationships between wind speeds, and the sizes of tropical vortices classified according to their cloud morphology. The sizes measured were the diameters of the central 'overcast circle' of dense, continuous cloud cover. Category 1 includes mostly tropical depressions, Categories 2 & 3 mostly tropical storms and Category 4 mature, classically formed hurricanes. *From Timchalk* (1965)

observations. The contingency table constructed to collate the results revealed that most of the estimates were within 15 kt of the actual wind speeds, encouragingly accurate bearing in mind the great wind speeds that were involved.

Many features of the tropical atmosphere that favour formation of hurricanes, such as increased low level vorticity inflow, above-normal warmth in the upper troposphere, and the existence of an anticyclonic circulation at some upper level over a low level cyclonic disturbance have been identified already (Riehl, 1954; Gentry, 1964). Meteorologists still know less, however, about the mechanics of formation of hurricanes than of any other phase of their existence. It is certain that some of the keys will be discovered as a result of satellite-based studies which involve both photographic and infra-red evidence, and which are processed climatologically to elucidate more clearly than hitherto the areas and conditions of hurricane genesis in various conventional data-remote regions of the world.

Leaving the intriguing topic of tropical vortices behind, and turning to extra-tropical depressions, there is, once again, much of interest to report. Meteorological satellites have provided photographs of cloud fields associated with an extremely large number of individual extra-tropical storm systems, and valuable analyses have been published by, amongst others, Boucher and Newcomb (1963), Widger (1964), Hadfield (1964) and Sherr and Rogers (1965).

Plate 3 has been reproduced in several meteorological works, but is worth reprinting again if only on account of its portrayal of occlusion and decay stages of extra-tropical vortices. It is singularly unfortunate that the second half of the life-cycle, while being that most frequently enacted over the British region, has also been that least appreciated until now.

Figure 10 is a new attempt to present a similar sequence of events in map form, incorporating patterns of isobars and wind flow at 1,000 and 500 mbs, and the distributions of the major bands or masses of clouds. Less organised cloudiness (for example, scattered cold sector cumulus and cumulo-nimbus in stage two, Figure 10b) has been omitted to avoid overloading the diagrams. The most prominent progressive development is the intensification of the spiralling of cloud and weather patterns in towards the centre as the life-cycle proceeds.

Figure 10 illustrates, at the 500 mb level, a mobile long-wave system involving a ridge-line moving away eastwards, followed by an incoming trough. These upper level changes promote an isallobaric zone at the 1,000 mb level, and where this coincides with a surface air-mass front (see Figure 10a), a vorticity centre is stimulated, marked by enhanced cumulus cloudiness (not shown) and a kink in the frontal wave of cloud. As pressure continues to fall at the surface, so the developing storm becomes larger and more intense, and the warm sector begins to be narrowed from the depression centre outwards as air spirals in towards the centre of vorticity (Figure 10c). The warm frontal cloudiness is distinguished by

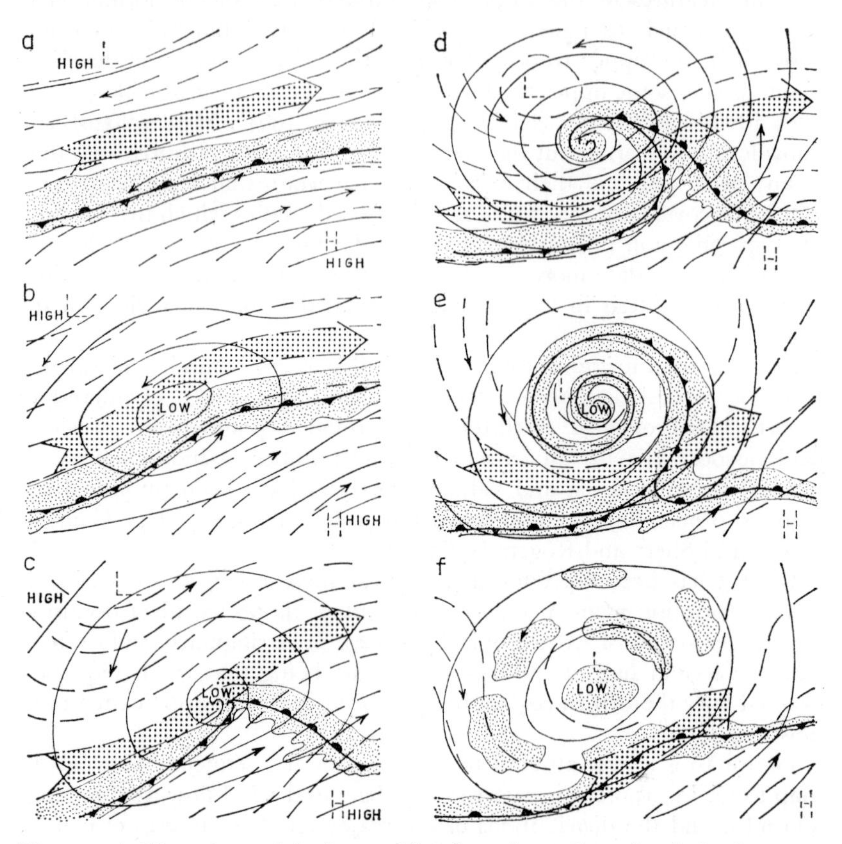

Fig. 10 A life-cycle model of a mobile, frontal, northern hemispheric extra-tropical cyclone. Closely stippled areas are major bands of sheet clouds. Broken lines and letters relate to 500 mbs, contours and winds, continuous lines to sea-level isobars. Open stippling indicates jet-stream cores.

longitudinal stripes of various degrees of brightness, while, behind the front, bands of cloud at right angles to the long axis broaden and thicken in the upglide zone where warm moist air is forced to rise over the cold. The cold front is characteristically mottled on satellite photographs, because of shadows cast by the taller cumulo-nimbus cloud tops that develop in this zone of greater instability. As occlusion begins, so a clear slot or groove of colder, drier air is drawn around the cloud disc that now marks the vorticity centre (Figure 10d), and there is a movement together of the upper and lower level pressure centres. As the joint processes of uplift of warm air and coiling inwards of cold air proceed, so the patterns of cloudiness and weather become distinctly concentric rather than radial. The energy advantage of the warm air is partly translated to kinetic energy—driving the circulation more rapidly at this stage than at any other—and partly shared with the cold air. The planes of contact between warm and cold air become more extensive as occlusion coiling proceeds, until, as in Fig 10f, the post-mature gently rotating circulation lies polewards from the main local air-mass front, and comprises largely homogenised air with temperature and humidity characteristics intermediate between those of the initial warm and cold sectors. In this way a meridional heat transfer is effected, in a polewards direction. Cloudiness and rainfall is broken and scattered, but still roughly concentric around the centre of rotation. Finally, it should be noted that, throughout the lifecycle as a whole, relatively warm air masses are marked by layered clouds (i.e. the cloud patterns illustrated in Figure 10), and cold air by clear conditions or cumuliform clouds (see Plate 3). Also the upper level jet-streams, associated with the surface fronts and their upwards, polewards-inclined, baroclinic zones, tend to be displaced progressively towards the equator as the depression matures. Hence, in Figure 10b, the main jet-stream axis is to the north of the vorticity and pressure centre, while in Figure 10e and 10f, the axes are well to the south.

The significance of these progressive changes in the organisation of depression weather in the climatology of north-western Europe has been recognised very inadequately in the past. In the summer season, relatively weak but youthful to mature systems tend to travel rapidly across the British Isles down the pressure gradient towards the continental thermal low pressure trough. In the winter, however, 'blocking highs' over the continent retard the movement of depressions in the mid North Atlantic, so that most depressions (with the occasional exception of a young secondary) arrive in the British region in mature or dissipating stages, giving, perhaps, no more rainfall than a summer depression, but causing more prolonged rainy periods that are difficult to forecast. Gloomy inversion conditions may develop as the old depressions fill, often over the British Isles.

Requirements of space in the present article do not permit the discussion of hurricanes or extra-tropical storms to be developed to greater length. Neither do they permit detailed discussion of satellite studies of other types of weather systems. In passing, however, mention may be made of satellite studies of the monsoon over India by Rao (1966), who suggested that the 'burst of the monsoon' over India is presaged by a sudden northwards bulge of the intertropical convergence zone to the southern tip of India early in May, while this zone shifts north over Africa more uniformly during the same period.

Mesoscale eddy patterns in the wakes of islands have been studied by Hubert and Krueger (1962), and these have shown that some eddies may be purely mechanical eddies produced by the island obstacles while others may be the result of inertial oscillation or inertial instability. These last systems are, of course, embedded in *anticyclonic* circulations, and although surface areas of anticyclonic flow are, on average, rather more extensive than areas of cyclonic vorticity, they have long been the Cinderellas of dynamic atmospheric studies. Since the great majority of satellite studies has focused on low pressure patterns there is relatively little that could have been reported on the subject of anticyclones. This is a pity when one remembers that anticyclones in tropical and polar latitudes form some of the most important quasi-permanent features of the general circulation, and in the field of synoptic studies there are great opportunities for pioneer studies of anticyclones. Certainly satellite photographic and nephanalysis indications suggest that their internal air-mass and air-stream variations are substantially greater and more frequent than had been recognised earlier, and it remains to be seen what effects these may have elsewhere—for example the effects of trade-wind anticyclone variations upon the development of extra-tropical depressions and upon weather variations in the equatorial low pressure belt.

V Regional climatic studies

1 Theoretical considerations

While recognition, identification, and interpretation of organised systems of cloudiness are basically important in most analytical studies of weather satellite data, the climatologist is, by definition, interested in applying his results differently from the meteorologist. Although few truly climatological applications of these data have been made yet, it is possible already to indicate something of the range and scope of the fields of applicability that are open for the immediate future.

Landsberg (1945) wrote that

> the reflection from land surfaces is probably the most important single factor in the atmospheric heat budget. All estimates of the heat transactions in the Earth's atmosphere . . . have to be based on *assumptions* about cloudiness, because reliable cloud observations are lacking in many parts of the globe.

Cloudiness is of special interest in this kind of context because:

1 Its presence serves to shield the underlying atmospheric layers, and the Earth's surface, from certain wave-bands of incident solar radiation, and increases the amounts reflected to space.
2 Clouds are efficient barriers to the transmission of longer wave energy radiations from the Earth towards space, and they re-radiate some of this energy back to Earth.
3 Clouds act as temporary thermal reservoirs, for they absorb a proportion of the energy incident upon them.

Adem (1964a) attempted to incorporate these considerations into a new thermodynamic model designed to specify and predict mean seasonal temperatures in the mid-troposphere and at the Earth's surface. In its initial form, Adem incorporated *cloud amount* as the only independent variable parameter, and several other workers (for example, Namias, 1960) have discussed observed relationships between anomalies in cloud cover and the Earth-surface expression of the atmospheric temperature field. Later tests have shown that changes of one-tenth in mean cloudiness prompts changes of several degrees in surface temperatures (Adem, 1964b), leading Clapp (1965) to conclude that knowledge of mean seasonal cloud cover will become useful not only in testing modelling assumptions, but also in simulating cloud amounts so that these quantities may be generated eventually within the model itself.

Even more recently Adem (1967) has suggested ways in which pressure, density and wind flow patterns might be predicted from the previously obtained patterns of mid-tropospheric and surface temperatures. To quote Adem, 'it is evident that the wind field is coupled with the temperature field, and with heat sources and sinks, and that it must therefore be generated within the model'. Much depends, however, upon 'a more realistic formula for the cloud cover' (than the amounts of cloudiness alone) and, in particular, more realistic indices should be introduced relating to cloud distributions in the vertical.

From these and allied approaches it appears that a new task for the satellite climatologist, contributing simultaneously to descriptive climatology and to applied meteorology, is to examine global cloudiness on a latitudinal and a seasonal basis, taking account of (1) the strength of cloud cover, (2) cloud thickness, (3) the dominant types of cloudiness interpreted either generically and/or genetically, and also (4) the mean vertical structure of the lower atmosphere. Prior to the arrival of dependable weather satellites, routine-prepared global maps of meteorological quantities averaged over months and seasons included neither cloud cover nor net radiation patterns. Routine maps of the kinds that are so important in general circulation and long-range weather forecasting researches were confined to fields of geopotential at various levels, together with their derivatives, such as geostrophic winds and the thicknesses of deep layers of the troposphere. The significances of studies of average cloud cover are at least three-fold (Barrett, 1968a):

1 As initial inventory studies to replace earlier maps, largely estimated, of average cloud cover.
2 In the allied roles of reconnaissance for satellite-based dynamic climatological research, and of contributions to studies of monthly, seasonal and annual climatic variability which, hopefully, may lead to a better understanding of climatic change, and
3 In programmes of meteorological model-building for descriptive and predictive purposes. Paramount amongst these programmes are those concerned with atmospheric energy sources and sinks.

The remainder of this section describes summarily the most fruitful methods of satellite data analyses with these ends in mind, and the results of one regional study for illustrative purposes.

2 Nephanalysis studies

Since nephanalyses indicate sub-synoptic scale areas of cloudiness classified in terms of percentage overcast, they can be used as the bases of mean cloud cover maps of chosen regions. This method was pioneered by Clapp (1964) who, from standard TIROS satellite data, prepared maps of mean monthly cloud cover over quasi-global zones between latitudes of about 60°N and S. Clapp chose a 5° grid network and averaged the daily nephanalysis cloud values around the globe for each 5° intersection of latitude and longitude. His results, and those of later workers, have shown that the earlier estimated maps of cloud cover were grossly over-simplified, particularly in tropical and oceanic areas. Comparative studies, relating nephanalysis cloud values, and near real-time conventional observations indicate that the former tend to overestimate the proportion of overcast in cloudy areas, and to underestimate it where little cloud is present, but, remembering the continuity and uniformity of the satellite coverage of the globe day by day, nephanalyses undoubtedly afford much better material for global and synoptic studies than do the data derived from surface stations. The level of resolution claimed for the nephanalyses by the United States Weather Bureau is of the order of 2°, and Godshall (1968), in a paper describing cloudiness in the east and central Pacific region, employed a 2° grid network as its foundation. In the vicinity of coasts, however, the geographical accuracy of nephanalyses has been much better than that for several years, and the modern computer-based satellite mosaics have increased the accuracy and dependability of them all.

Of late, it has been suggested that nephanalyses may be useful base data for the compilation of maps of estimated monthly rainfall. (Barrett 1970.) The following equation was developed theoretically:

$$K_r = \frac{C\,\Sigma\,(Mp_1i_1c_1 + Np_2i_2c_2 \ldots Rp_6i_6c_6)}{100} \qquad (1)$$

where C is the mean monthly percentage of cloud cover; Σ is the sum of available daily nephanalysis cloud data in one calendar month; $p_1 \ldots p_6$ are assigned probabilities of rainfall for six states of the sky shown on nephanalyses; $i_1 \ldots i_6$ are assigned intensities of the same; $c_1 \ldots c_6$ are the six states of the sky; and $M \ldots R$ are the numbers of occurrences of $c_1 \ldots c_6$.

The rainfall coefficient K_r was evaluated for selected rainfall stations in the Australian region, and the results graphed for equivalent recorded rainfall totals.

The best fit regression curve was computed to relate the two and used to translate into rainfall estimates evaluations of equation (1) applied to grid intersections across the region from 90°–180°E and from 15°N–30°S.

This kind of method may make possible studies of rainfall patterns even in those areas whence conventional data are infrequent or unavailable. Apart from the improvement of atlas maps of global rainfall distributions, the ability to assess rainfall uniformly across both land and sea could be invaluable in global modelling of the world water balance. Currently global patterns of mean precipitation are compiled from surface data where these are available and from generalised meteorological assumptions where no such data exist.

3 Photographic studies

The most basic of all these studies was designed to utilise the raw digital brightness values which comprise the photographic radio signals from the satellites to the tracking stations. Arking (1964) employed a computer-based programme to select systematically brightness values from the tape recordings of photographic data. These values were then grouped and averaged for selected regions, and frequencies of cloud cover derived. Although the basic choice, namely at what digital level the presence of cloud is usually first indicated, was a difficult one, this type of approach will probably be developed in the future on account of its strong dependence on electronic computers.

The second type of study employs satellite photographs to investigate the preferred tracks of different classes of weather systems, and their most frequent areas of genesis, growth and decay—for example, the hurricane studies conducted by the United States Weather Bureau since the first weather satellites were launched in 1960.

The third, and potentially perhaps the most interesting, group of climatological studies depends upon the apparently most usual relationships between meso- and macro-scale striations and bandings in cloud fields and the directions and strengths of lower level wind flows. Disregarding here the various relationships that have been discovered in the upper troposphere (summarised in the section on jet-stream clouds in Barrett (1967b), and in the comprehensive paper by Johnson (1966) on cirrus clouds and upper tropospheric motion), the relationships between varieties of cumuliform clouds and geostrophic and observed wind fields near the Earth's surface have been considered by several workers.

It seems that, excluding the broad groups of topographically disturbed cloud types (see, for example, Conover, 1964; Fritz, 1965) and the taller cumulo-congestus and cumulo-nimbus clouds which must be interpreted in terms of the vertical wind shear through the cloudy layer (see Ericksson, 1964), the remaining types of clouds frequently become aligned along the direction of air flow. Kuettner (1959) has pointed out that, in the organisation of clouds in rows down to the meso-scale, from the tropics to as far north as the Arctic Circle, 'most cloud bands line up in the direction of flow, and originate in convective layers.' This view has received empirical support from Gaby (1967), who studied comparatively weather satellite photographs and the relevent conventional weather maps. Gaby concluded that in low latitudes at least the more prominent cumulus cloud lines, constituting the locally 'dominant cloud mode', commonly lie parallel to the directions of surface winds.

Although less work has been done on the preferred alignments of bandings and striations in more continuous cloud fields, the paper by Lyons and Fujita (1968) is a significant pioneer, where the alignments of meso-scale striations in oceanic stratus sheets are examined across the North Pacific. Previous laboratory experiments by Faller (1965) had successfully simulated parallel cloud bands in a rotating tank, and Faller concluded that in continuous or near-continuous cloud fields the bandings should be aligned usually to the left of the geostrophic wind in the northern hemisphere, at angles between 10–17°, with an average of 14°. The findings of Lyons and Fujita are closely confirmatory in that the stratiform cloud rolls which they examined on satellite photographs generally lay across the isobars at about 16°. This angle, it can be suggested, is very significant since it approximates to the mean angle at which lower level wind flow crosses isobars from high to low pressure near moderately smooth surfaces in mid-latitudes.

Finally, the results obtained by the present author (Barrett, 1986b) suggest that the shallow layers of strato-cumulus beneath the trade-wind inversions of tropical and sub-tropical ocean areas also became banded most frequently in a down-wind direction, and that acceptable streamlines can be derived from the long axes of the cloud bands.

These considerations therefore permit the inference of streamline patterns in the surface layer of the atmosphere provided that care is taken to use only the shallow clouds for these purposes. Otherwise the effect of shearing in the vertical serves to contaminate the resulting pattern. Since there are many terms for the various elongated arrangements of clouds (for example, cumulus 'streets' or 'rows', cirrus 'streamers' or 'fronds', stratiform 'striations' or 'rolls', strato-cumuliform 'stripes' or 'bands') it is proposed that linear features of all cloud types at the meso- and macroscales be referred to as *nephlines*. Depending on the cloud genera involved, one may speak therefore of cumulus nephlines, and of stratus and stratocumulus nephlines. The preliminary tracings of selected shallow nephlines from satellite photographs become 'nephline analyses'. Generalising these cloud axes, the resulting 'smoothed nephline analyses' serve as acceptable streamline substitutes for use in studies of vorticity patterns and distributions and as bases from which isobaric patterns may be drawn if required. It is clear that modifications of cloud fields by land topography restrict the reliability of this approach to oceanic areas, but the potential contributions to climatological knowledge are great none the less; seven tenths of the Earth's surface is oceanic, and these areas are the least known from climatological standpoints.

4 Central America and the tropical eastern Pacific: a regional case study in satellite climatology

Crowe (1949), writing about the tropical eastern Pacific, described it as 'the most empty ocean of the world. The few island stations in this vast area with a record of more than a few years lie, without exception, towards the margins of the (major pressure) systems.' Hence, climatologically, much remains to be learnt about the region. Although satellite data can be evaluated most fully when corresponding conventional data are available too, they can, however, be of great value interpreted alone.

In recent years, several workers have attempted to identify and track the movements of the so-called inter-tropical convergence zone (ITCZ), through considerations of wind flow (particularly its speed and constancy) (Crowe, 1950, 1951), streamline patterns (Palmer, 1951), and wind parameters and percentages of cloudiness (Godshall, 1968). The study summarised below seeks to portray mean cloud patterns in and around Central America from nephanalysis indications; it seeks to depict common synoptic distributions of cloudiness and wind flow derived from satellite photomosaics, and to trace the daily and seasonal changes in the alignment of the ITCZ, or *equatorial (low pressure) instability axis*. In view of the now more obviously complex nature of the low pressure belt and the various definitions of the ITCZ, *equatorial instability axis* is a term to be preferred as far as the near-equatorial patterns of deep cloudiness and instability weather are concerned.

The photo-mosaics were constructed from individual frames taken by NIMBUS II during its operational life-time from 18 May to 31 August, and were especially valuable for detailed studies on account of their good resolutions, which permitted the identification of features about 1½–2 km across. The nephanalyses were compiled from a variety of satellite sources, being based on NIMBUS II, ESSA IV and ESSA V photographs. Satellite data indications were compared with conventional data as far as possible; but even the best synoptic maps for the region compiled by the Venezuelan Weather Bureau and the Howard Air Force Base (United States Air Force, Panama), only provided a partial coverage, and the patterns they portrayed were obviously tentative away from the Central American isthmus. The photographic coverage, after swaths of poor resolution and poor illumination had been removed, represented a 60% sample of the days from mid-May to late August (see Figure 24).

The results of the nephanalysis studies of average cloudiness are portrayed by Figures 11–18. Figure 11 shows that in this summer season in the northern hemisphere a broad belt of relatively strong overcast was aligned latitudinally with its axis at about 7½°N. Similar belts of strong cloudiness extended north to the west of 115°W, and south between the South American coast and about 102°W. Areas of notably low cloudiness included the Peruvian coastlands, the Gulf of California and its immediate land areas, and the west central Caribbean.

The fortnightly maps (Figures 12–18) help to illustrate the progressive changes that took place in the distributions of the various air masses and air streams with which the cloud fields were associated. In late May (Figure 12) the alignment of the main east–west cloud axis was roughly latitudinal at about 6°N; by the first half of June its position had altered little, but there was a general strengthening of cloudiness in the cloudy regions and a tendency for the areas of sparse cloudiness to become more marked. In the second half of June the cloud axis associated with the equatorial low pressure belt bulged northwards to the west of the isthmus, adopting a curvilinear form. Strong cloudiness developed along the isthmus from Yucatán to the north-west corner of South America, and the western American contrasts between coastlands and oceans remained sharp. In early July cloudiness along the isthmus was considerably stronger than over the sea areas to the east and to the west, and the main bulge of the curvilinear equatorial low pressure cloudiness was pushed away westwards from the isthmus. Later the same month this cloudiness retreated

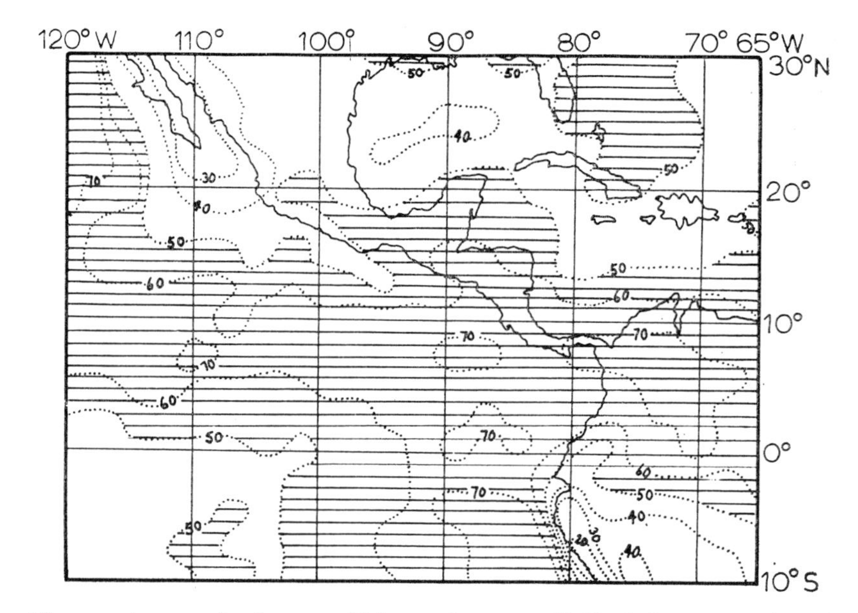

Fig. 11 Average cloud cover, 16 May–31 August, 1966. Shaded areas experienced 50% cloud cover.

Figs. 11–18 Maps of average cloud cover, in percentages, covering the eastern tropical Pacific Ocean and the Central American region, derived from nephanalysis evidence. (*From Barrett* (1970*b*)

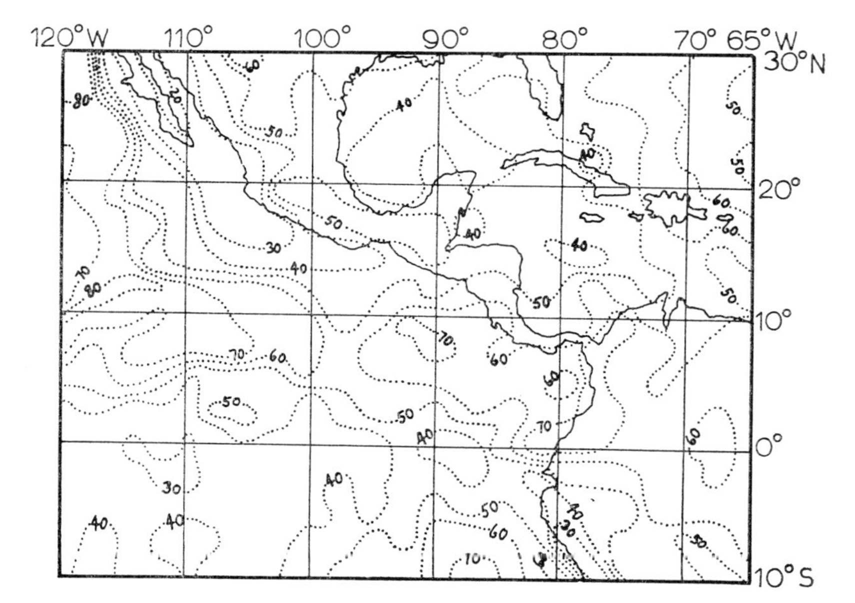

Fig. 12 Average cloud cover, 16 May–31 May, 1966.

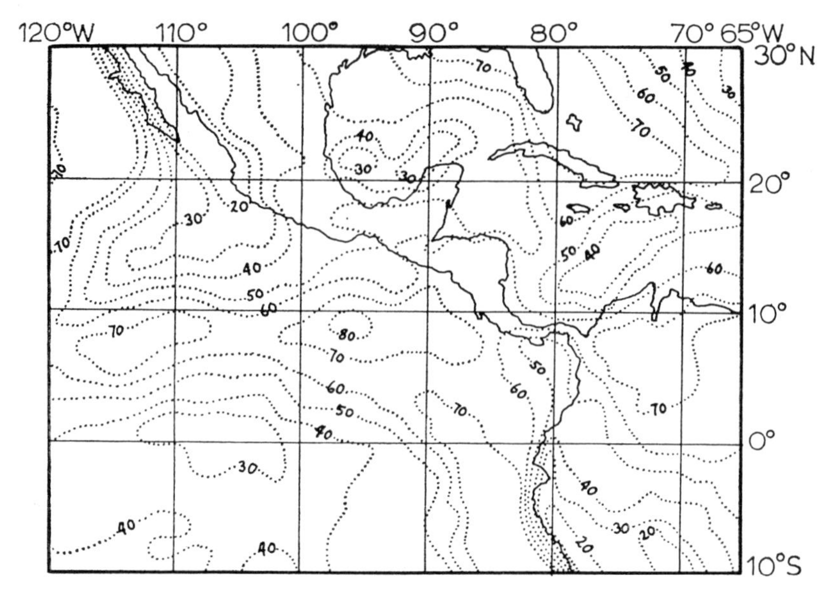

Fig. 13 Average cloud cover, 1 June–15 June, 1966.

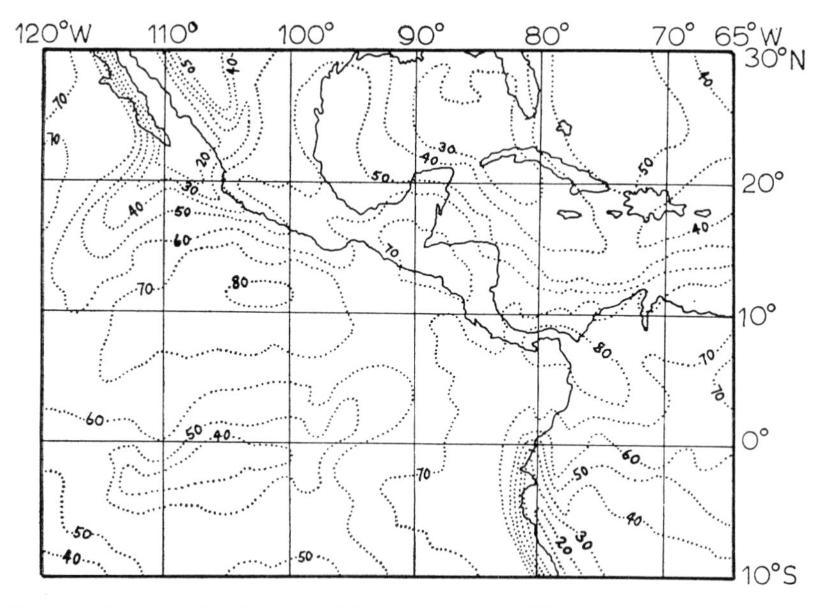

Fig. 14 Average cloud cover, 16 June–30 June, 1966.

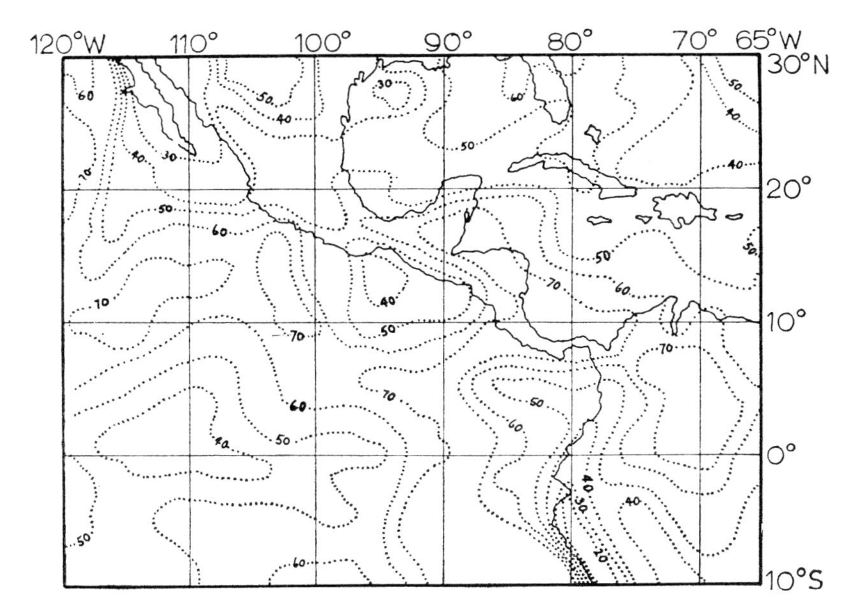

Fig. 15 Average cloud cover, 1 July–15 July, 1966.

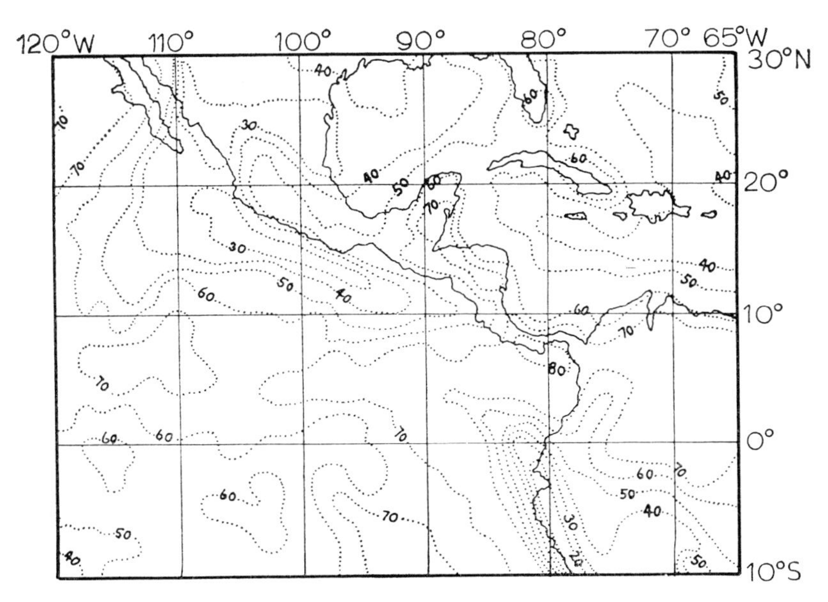

Fig. 16 Average cloud cover, 16 July–31 July, 1966.

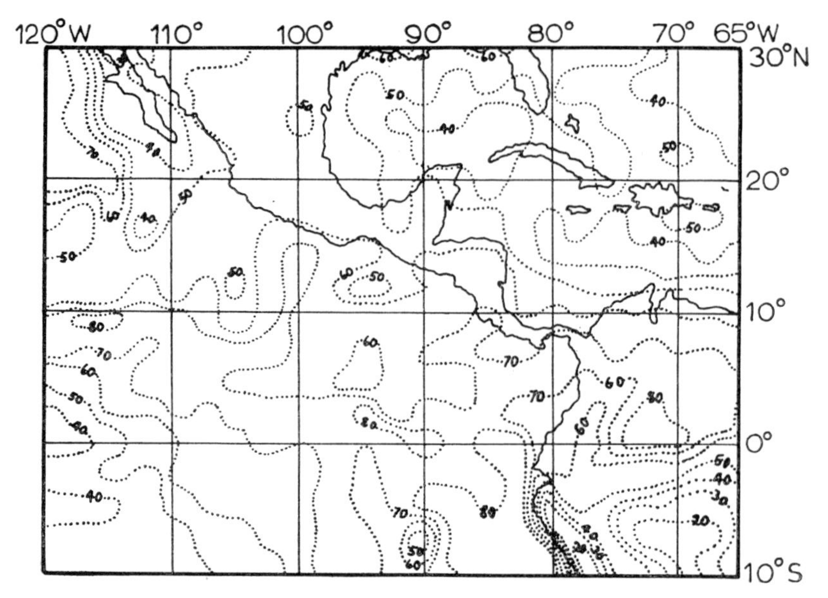

Fig. 17 Average cloud cover, 1 August–15 August, 1966.

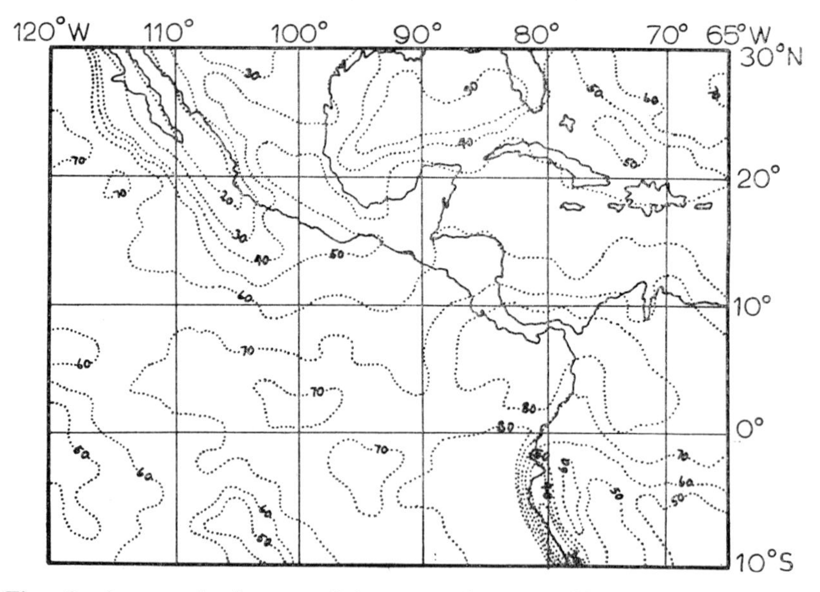

Fig. 18 Average cloud cover, 16 August–31 August, 1966.

southwards to lie along a more nearly latitudinal axis, once again, between about 0–10°N. By August, the cloud belt had broken up, and Figure 17 portrays a more complex pattern of relatively cloudy and relatively cloud-free areas than any other map in this series, though the coastal contrasts in the Californian and Peruvian regions still remain strong. Finally, in late August, there was a pattern more similar to that in late May than in any other period, a latitudinally aligned belt having re-formed over the eastern equatorial Pacific region, with its northern margins around 8°N.

These cloud maps, of course, contain a wealth of other interesting features that cannot be discussed here; suffice it to say that they are the most detailed cloud maps yet prepared for this region of the world.

From the satellite photographs, and the dependent nephline and smoothed nephline analyses it was possible to examine and interpret the cloud distributions and periodic fluctuations much more closely, In all, five distinct synoptic arrangements of cloudiness and weather systems were identified, although hybrids also occurred. These hybrids were anticipated within the continuously changing fluid medium of the atmosphere. Plates 4 and 5 exemplify the NIMBUS II mosaics upon which this part of the study of Central America was based. Further illustrations, and a more detailed discussion of the whole project and its results appear elsewhere (Barrett, 1970b).

Plate 4 represents one of the best of the mosaics, and is particularly useful in that it contains many well-developed synoptic meteorological features characteristic of the region in question. These include:

1 Trade-wind anticyclonic cloudiness in the South Pacific. The prominently banded pattern includes both open and closed modes of strato-cumulus cloud cells, and also areas of ('actinoform') cloud discs, vermiculate (irregular worm-like), elliptical, and blown-out strato-cumuliform patterns.
2 Strong cloudiness off the South American coast, contrasting markedly with the clear skies over the desert coastlands of South America south of the Gulf of Guayaquil. The strong cloudiness is of a strato-cumulus species, and is associated with the cold ocean current flowing equatorwards along the coast. On-shore winds are heated by the land, and their humidity falls dramatically. Little cloud is seen over the Andes. Lake Titicaca is clearly visible, whilst patches of amorphous, low level stratiform cloud overlay the marshy areas of Salan de Uyant and Salan de Coipasa, around 20°S, 67°W.
3 To the east of the Andes the cloudiness consists of bright, amorphous patches, relating to the convective outbreaks that develop diurnally within the broad low pressure area over the large equatorial air mass.

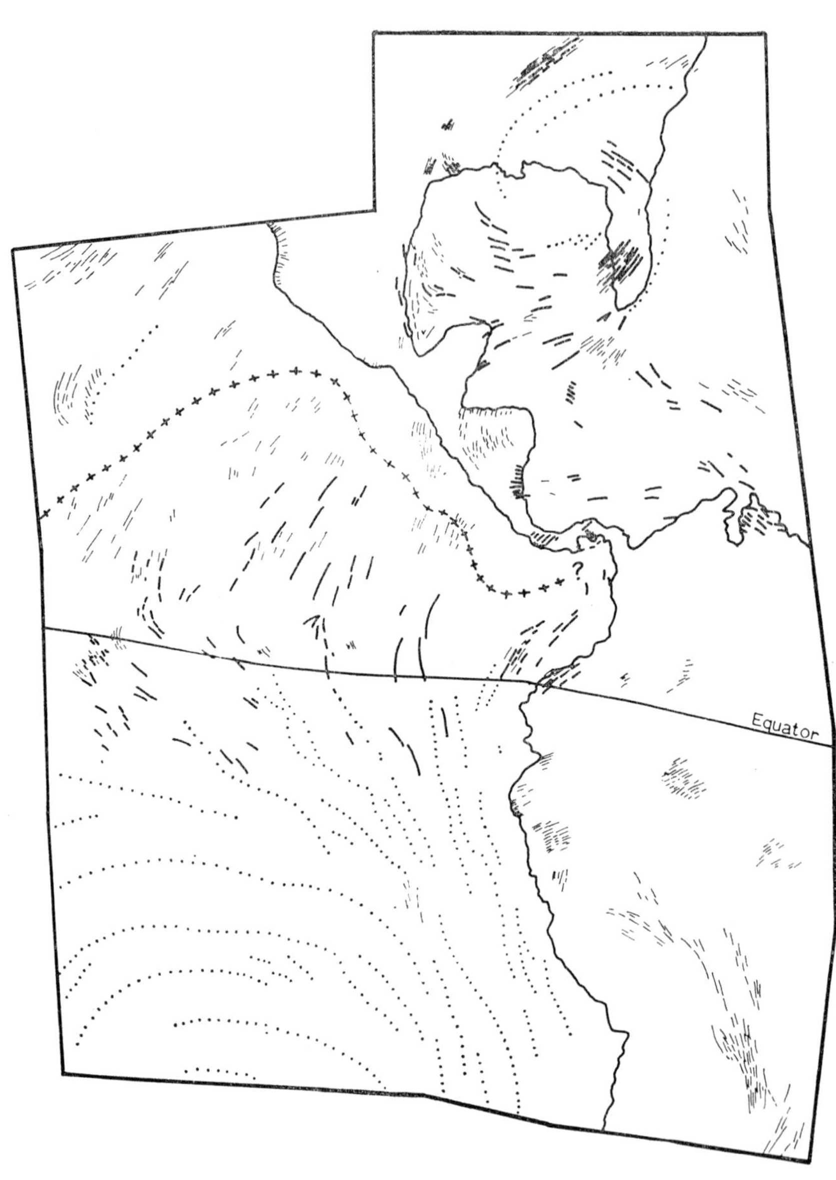

Fig. 19 Nephline analysis of Plate 4. The low pressure instability axis is shown by the line of crosses. Cumuliform nephlines are shown by continuous lines, whilst strato-cumuliform nephlines are dotted. Similar symbols are employed in Fig. 21. *From Barrett* (1970*b*)

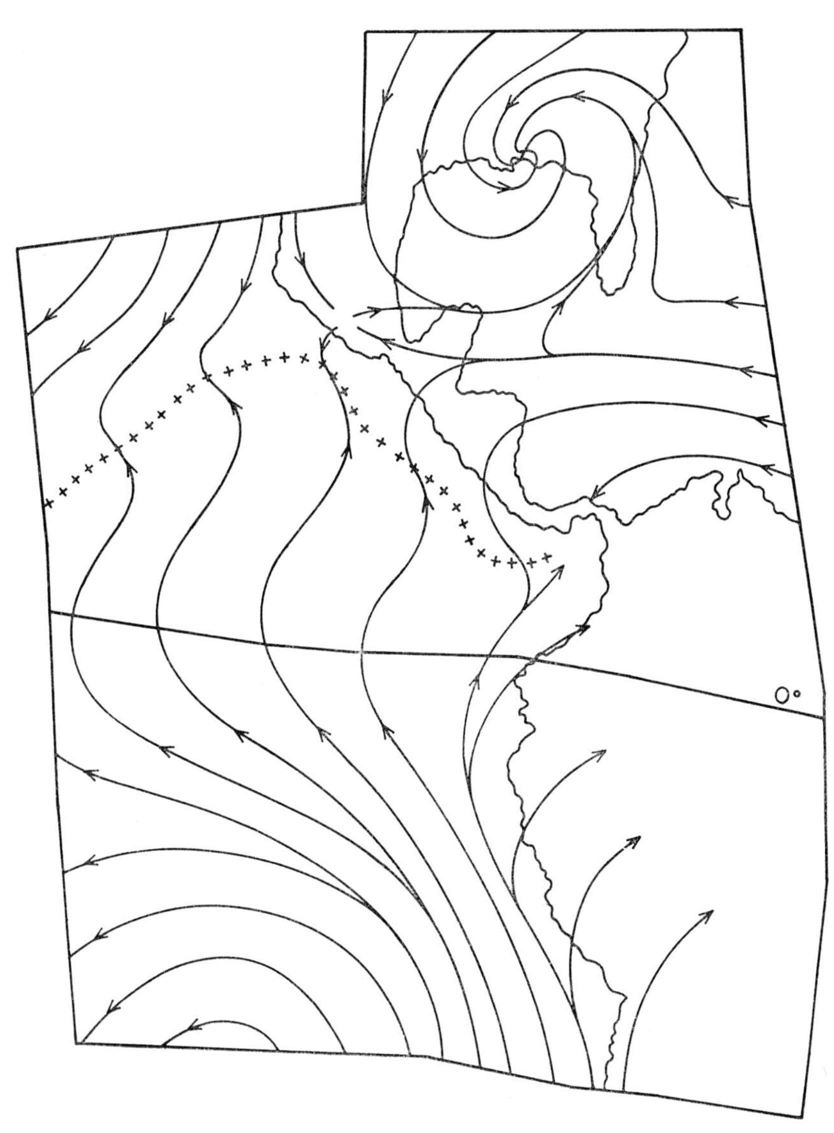

Fig. 20 Smoothed nephlines, approximating to streamline patterns, drawn from the evidence of Plate 4 and Fig. 19.

4 The Caribbean is mostly cloud-free, under the influence of easterly anticyclonic air flow, but
5 Over Florida there is the fully mature, beautifully proportioned tropical cyclone ALMA.
6 To the west of the Central American isthmus the zone of instability cloudiness is well marked, adopting a curvilinear form parallel to the isthmus to the east, and merging with overland convective cloudiness near its junction with South America.
7 In the north-west corner, fine lines of cumulus and strato-cumulus appear in the North Pacific anticyclonic region. These clouds give way to some cumulo-nimbus lines towards the major instability zone.
8 Suppressed cloudiness occurs in a broad zone of subsidence along the equator south of the instability axis.

Figures 19 and 20 portray the flow patterns that were derived from the larger original mosaic. They also show the position of the low pressure instability axis deduced from these patterns, and the patterns of best developed instability clouds. This example should be compared with Figure 13.

Plate 5, illustrating a second type of synoptic pattern, portrays conditions on 24 July, 1966. In this case it is not possible to identify a linear instability belt west of Central America, where, instead, there appear a number of amorphous patches of cloudiness, which, by their size, brightness and indistinct margins, must be interpreted as tropical disturbances. These contribute to the kind of cloud distribution illustrated by Figure 16, and early indications are that in the wedge-shaped area of low pressure west of Central America, disturbances such as these tend to develop and dissipate very rapidly unless they grow to become upgraded to the tropical depressions, tropical storms, and even hurricanes, that characterise the region in late summer and early autumn. To the east of the isthmus, a fine example of an easterly wave configuration is apparent in the cloud field: this is more graphically displayed by Figures 21 and 22. Meanwhile, over the isthmus itself further centres of strong instability cloudiness appear.

Since the five recognised synoptic types were differentiated from one another most importantly in terms of their patterns west of Central America, they can be summarised as follows:

Type 1 The equatorial instability axis was relatively straight, and latitudinally aligned, about 7½° north of the equator.

Type 2 The axis was highly curvilinear, the main thrust of ex-southern hemisphere air being northwards a few degrees west of the isthmus (see Plate 4 and Figures 13, 19 and 20).

Type 3 The main instability axis lay along the isthmus, and was marked

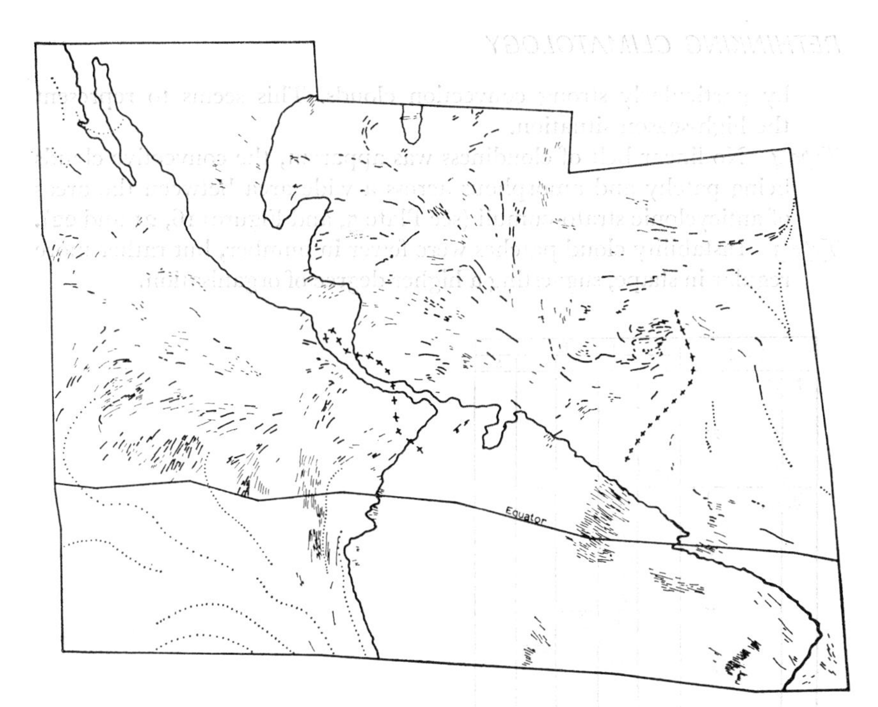

Fig. 21 Nephline analysis of Plate 5. *From Barrett* (1970*b*)

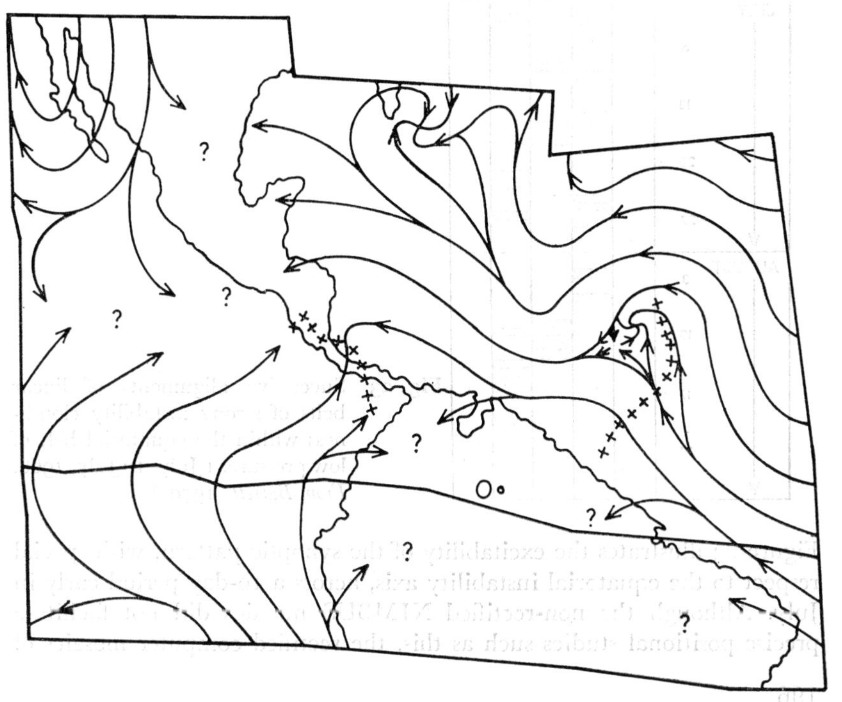

Fig. 22 Smoothed nephlines constructed from the cloud evidence of Plate 5 and Fig. 21.

by particularly strong convection clouds. This seems to represent the high-season situation.

Type 4 No linear belt of cloudiness was apparent, the convective clouds being patchy and amorphous across a wide area between the areas of anticyclonic strato-cumuli (see Plate 5, and Figures 16, 21 and 22).

Type 5 Instability cloud patches were fewer in number, but rather more regular in shape, suggesting a higher degree of organisation.

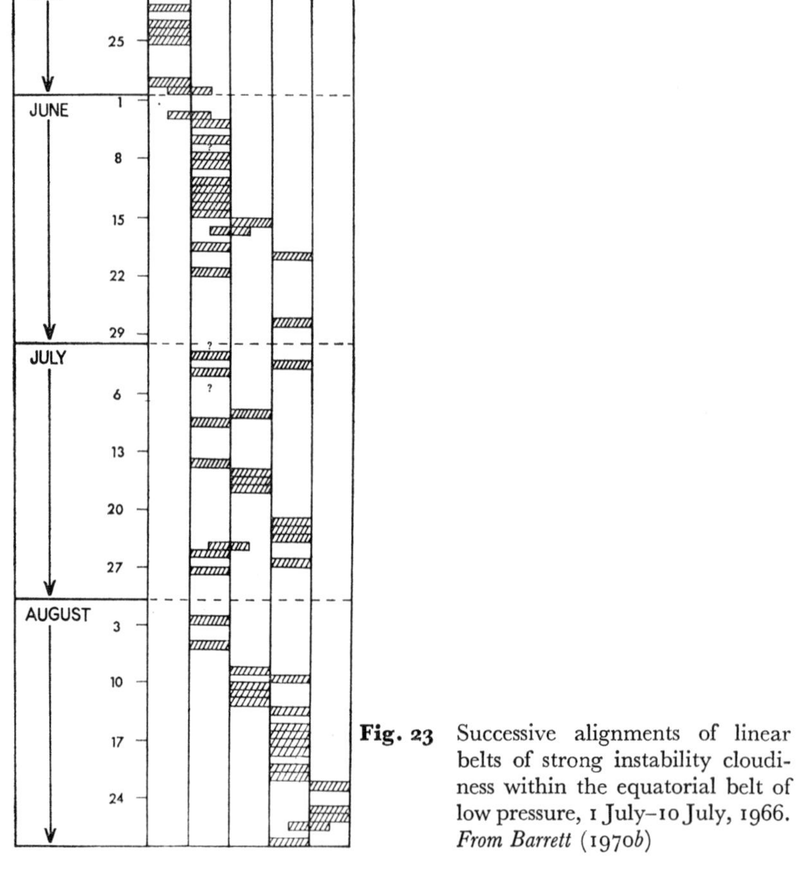

Fig. 23 Successive alignments of linear belts of strong instability cloudiness within the equatorial belt of low pressure, 1 July–10 July, 1966. *From Barrett* (1970*b*)

Figure 23 illustrates the excitability of the synoptic pattern, with special respect to the equatorial instability axis, across a 10-day period early in July. Although the non-rectified NIMBUS mosaics did not facilitate precise positional studies such as this, the rectified computer mosaics of

ESSA satellite photographs will assist the student of daily weather fluctuations in any similar attempts.

Figure 24 is a tabular summary of the observed succession of daily synoptic types from mid-May to late August. Although the fluctuations

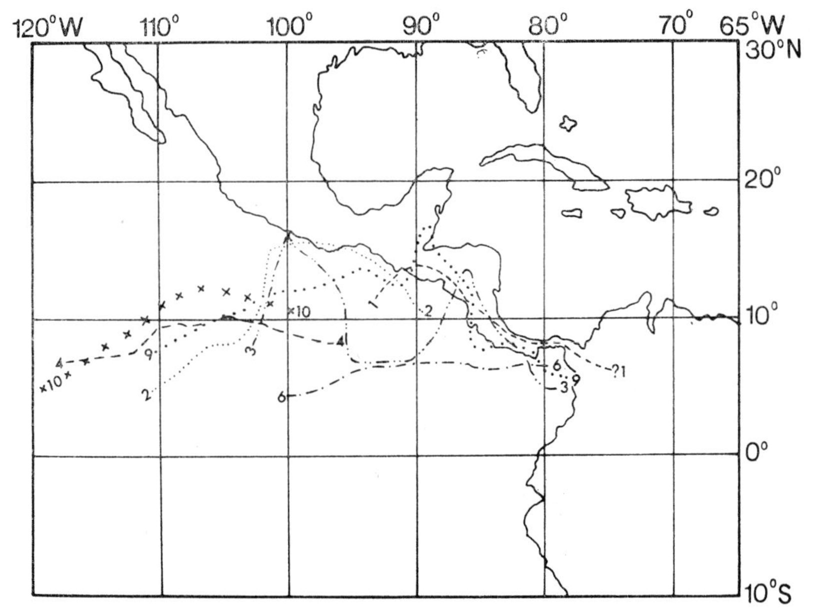

Fig. 24 The occurrence of the five observed synoptic patterns in the eastern tropical Pacific Ocean and the Central American region, 18 May–31 August, 1966. *From Barrett* (1970*b*)

again speak of high excitability, the progression from a dominance of Type 1 to a final dominance of Type 5 is noteworthy, and helps to explain the cloud maps (Figures 13–19). Considering all this evidence it seems hardly surprising that the ocean area to the west of Central America could be described by Crowe (1951) as one of light and variable winds: there are many factors—some extra-terrestrial, some upper tropospheric, some lower tropospheric, and the remainder surface factors—which all play and interplay, in the summer season at least, along this inter-hemispheric frontier zone.

5 Conclusion

It may be concluded that methodologies developed to invoke the evidences of satellite photographs and nephanalyses in problems in dynamic climatology can yield some useful results over and above those previously

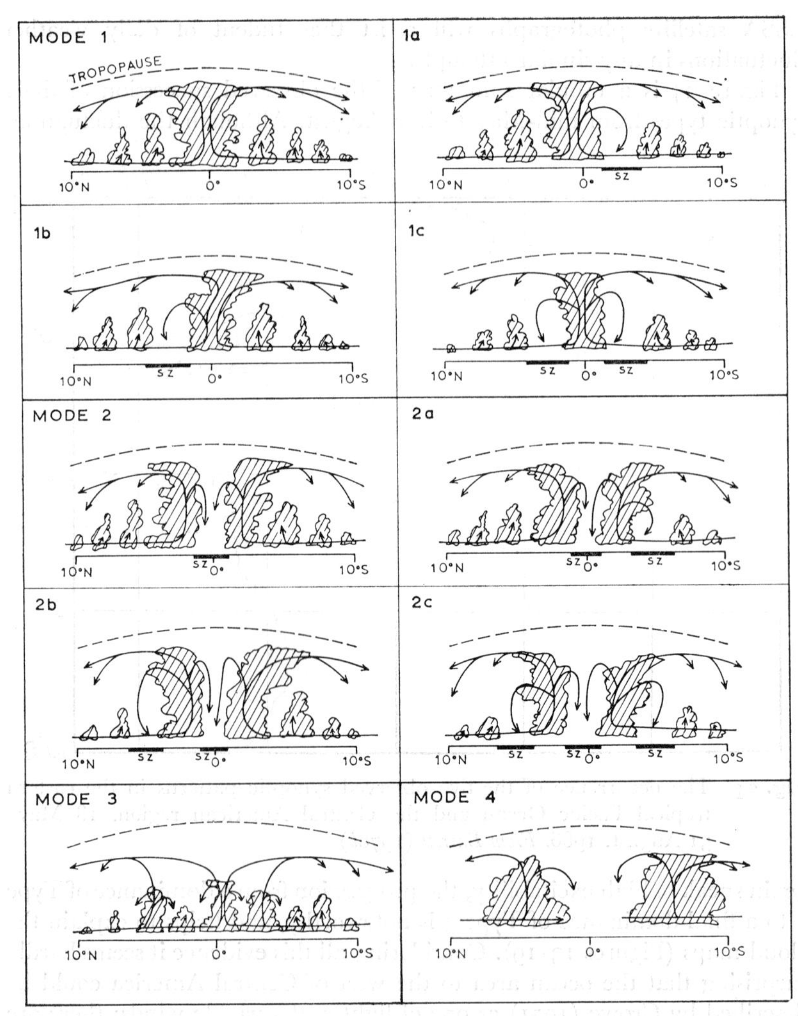

Fig. 25 Idealised meridional sections through the inter-tropical convergence zone, illustrating the possible range of forms that the equatorial instability axis may adopt. The classification is based jointly on the cloud organisations and subsidence zones (S.Z.) that develop roughly parallel to the main instability belts. The main distinctions are, therefore, between arrangements involving single instability axes, double instability axes, shear-line axes with subdued instability cloudiness, and patchy instability cloudiness lacking linear organisations.

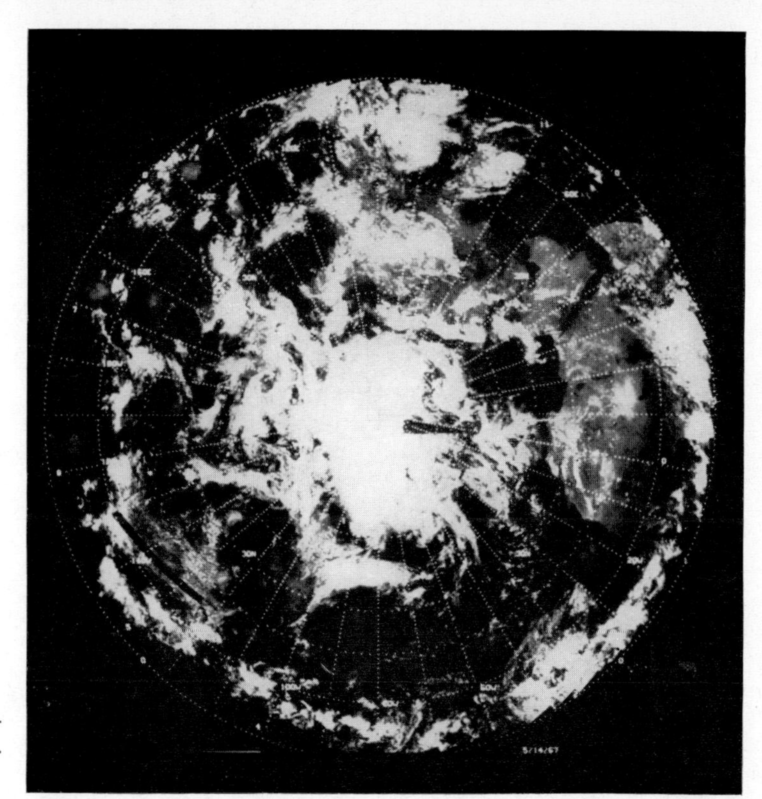

Plate 1 A computer-rectified, brightness-normalised, polar stereographic mosaic for the northern hemisphere, 14 May, 1967. (*ESSA photograph*)

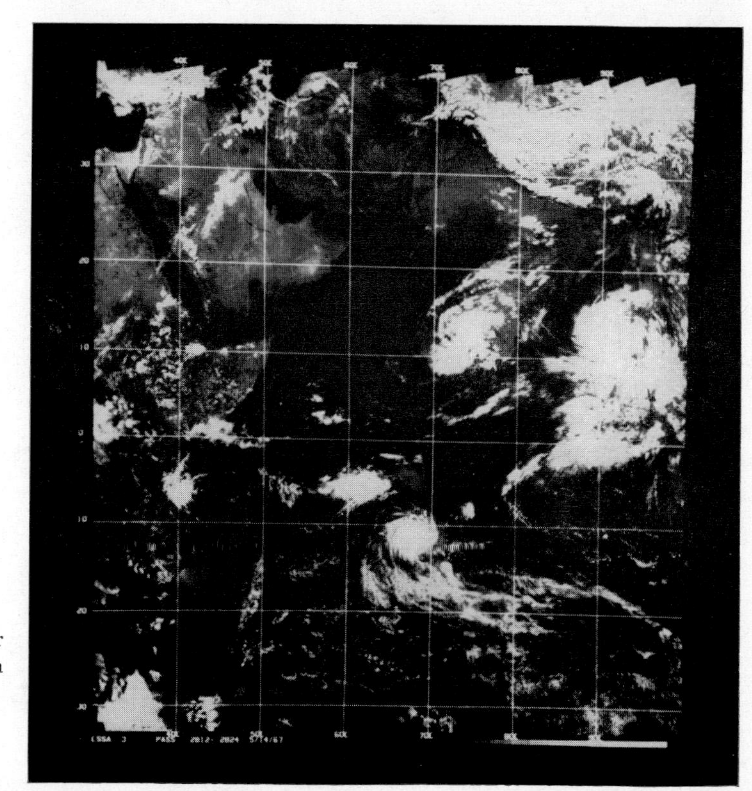

Plate 2 A computer-rectified, brightness-normalised, Mercator mosaic for the western Indian Ocean, 14 May, 1967. (*ESSA photograph*)

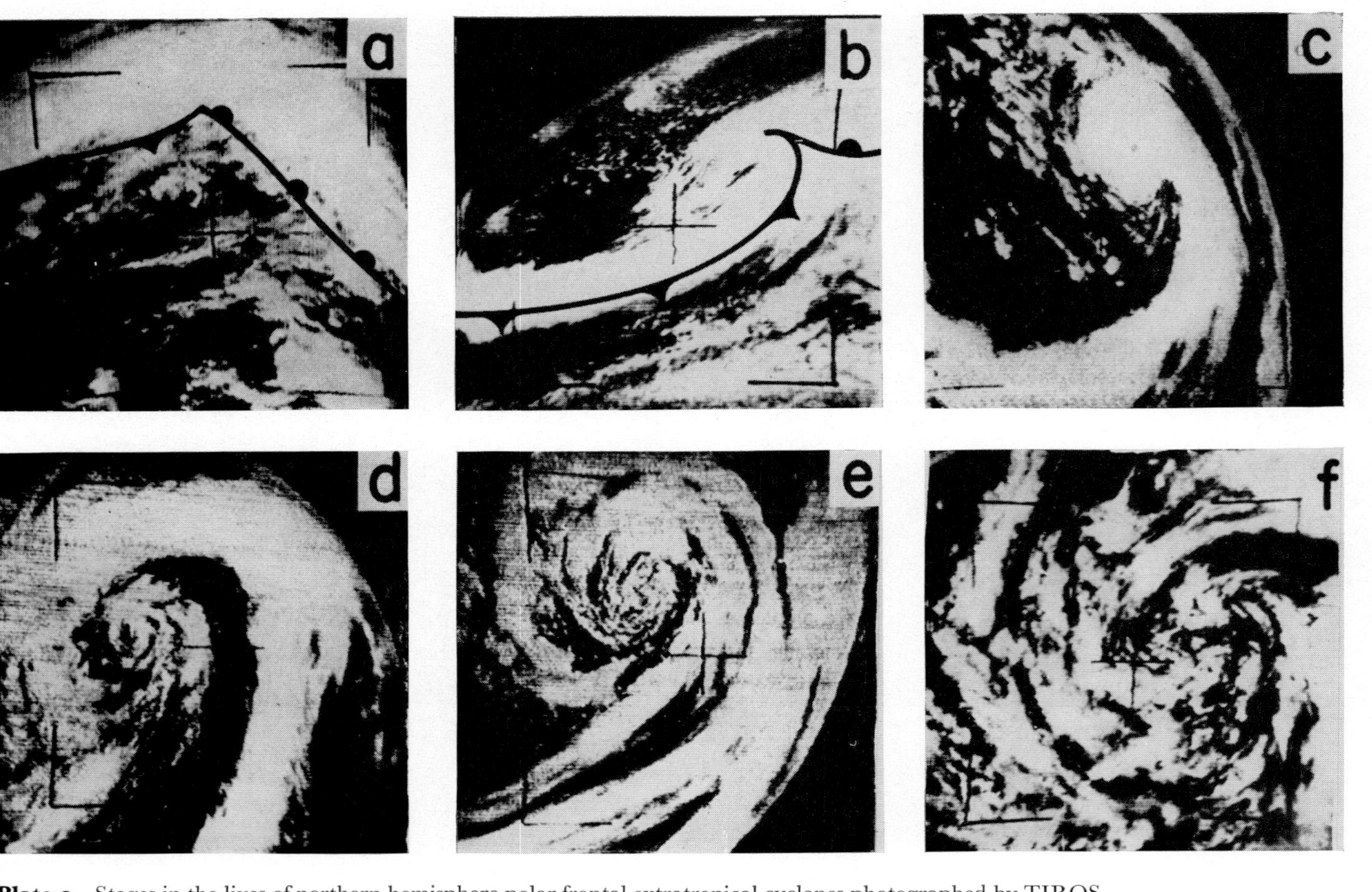

Plate 3 Stages in the lives of northern hemisphere polar frontal extratropical cyclones photographed by TIROS satellites. Similar stages are illustrated diagrammatically in Fig. 10. (*ESSA photograph*)

Plate 4 NIMBUS II mosaic, central American region, 11 June, 1966, compiled in the University of Bristol from original contact prints supplied by NASA. Either of the two component swaths comprises three rows of photographs taken by the fan-like array of cameras. The two swaths were obtained from successive satellite orbits about 100 minutes apart. The marker spot represents the location of the Panama Canal Zone. (See Figs. 19 and 20.)

Plate 5 NIMBUS II mosaic, central American and Caribbean regions, 24 July, 1966, compiled in the University of Bristol from original contact prints supplied by NASA. Note especially the fine easterly wave cloud pattern in the east. (See Figs. 21 and 22.)

obtained by other workers in similar areas using conventional data alone. The most important possibility is that lower tropospheric streamlines may be derived from satellite photographs, and constitute highly valuable documents in previously data-remote regions, especially oceanic regions. Nephanalyses are useful in reconnaissance and cloud inventory capacities, and it should be possible to develop computers and computer programs to alleviate the routine calculations. In particular, early meteorological satellite investigations have suggested that the structure and variability of cloudiness, and the associated patterns of circulation and weather within the equatorial low pressure trough are considerably more complex than earlier, conventionally based studies indicated. Further circulation forms, other than those suggested in section 3 (above) are indicated by multiple exposure averages of satellite mosaics compiled by Kornfield *et al.* (1967), in part explained by Bjerknes (1969).

Figure 25 attempts to model the range of circulation patterns indicated by satellite photographs of equatorial oceanic regions. Each section represents an idealised arrangement of strong convective cloudiness across the equator. The subsidence zones, each linked with a localised 'Hadley cell within a Hadley cell', are of particular interest. Studies of their variable patterns and distributions around the globe across time should be carried out to assess their relative significances within the tropical circulation as a whole. Recent work by Bjerknes (1969) suggests that patterns of the Mode 2 family in Figure 25 may be related to zonal ('Walker') circulations in the tropical Eastern Pacific. Certainly there can be no doubt that atmospheric circulations over tropical oceans are much more complex and variable than classic, Hadley cell models suggested, involving exclusively meridional circulations.

VI Satellite data in the perspective of dynamic climatology

Satellite data can be used to examine and illustrate many features of the Earth's atmosphere, as this paper has sought to show. In what ways may the data be employed to greatest effect in dynamic climatology? Whilst this question is not easy to answer, it can be argued that satellite photographs may promote new studies of vorticity distributions, nephanalyses will assist students of mean cloud cover, and infra-red measurements will throw light upon the important energy relationships that exist within the

Earth/atmosphere system and between this and the primary source of fuel for the atmospheric motor, namely the sun.

The streamline patterns derived from satellite photographs and illustrated in section V could facilitate the identification of the main low pressure belt in the vicinity of the equator in terms of the curvature of the streamlines—whether this is positively or negatively cyclonic. Infra-red data, on the other hand, although not discussed in detail here, have been shown to be potentially useful in the delimitation of various radiation and temperature regimes, including the distributions of world regions of net radiation gain and net radiation loss, by considering the energy fluxes entering and leaving the top of the atmosphere. It is quite possible that photographic and infra-red data, analysed in these ways, might contribute much to new approaches to the study of the general circulation of the atmosphere, and hopefully, to climatic classification.

Although in the past pressure, temperature, rainfall, etc. have been the basic elements considered in these related fields of study, it might prove more informative and helpful to consider qualities of the atmosphere that appear to classify themselves more naturally. Two of the greatest problems in climatic classification in the past have stemmed from the fluid continuity of the atmosphere in both space and time, and the extreme difficulty of quantifying subregional boundary values within this continuous medium. The present author has suggested (Barrett, 1968b) that it might be worthwhile to consider instead the patterns of vorticity, net radiation loss or gain, and the distribution of the precipitation/evaporation ratio to measure atmospheric humidity, since these three parameters can all be examined on the basis of yes/no types of questions, and yield distribution patterns which contain higher degrees of physically determined differentiation. In the case of vorticity, it can be argued that distributions of positive and negative vorticity afford much more reliable and meaningful subdivisions of the atmosphere than 'high' or 'low' pressure, since it is a common observational fact that some pressure values may occur quite commonly either in an anticyclonic or a cyclonic circulation. In the case of radiation patterns, areas of net heat gain and net heat loss are much more significant in dynamic climatology with its interest in heat sources and sinks than the often arbitrarily subdivided temperature fields examined in descriptive climatology. Thirdly, the wetness or dryness of the atmosphere itself is better measured by the precipitation/evaporation ratio than by rainfall totals, and it is possible that satellite infra-red measurements may lend themselves to those types of humidity studies.

It has been concluded on theoretical grounds that combining the patterns derived from the mapping of answers to these three yes/no questions on an annual basis alone would give a maximum of eight possible combinations. This number rises to 64, on a seasonal (half-yearly) basis. Of these 64 about half may be expected to occur in reality. Thus the basis for a new detailed classification scheme could be laid, and by considering, for example, monthly patterns rather than annual patterns, very complex systems of subregions would emerge. Their variations from season to season or year to year might throw important light on the patterns and mechanisms of 'short-lived deviations of the general circulation' and contribute to the current rethinking of climatology along more physical lines.

This type of study would, of course, involve the use of both satellite and conventional data, using either to fill the gaps left by the other, and might, most hopefully, finally feed useful information into applied meteorology through the gateway of long- and extended-range forecasting. It is good, however, to conclude on a note of caution: weather satellites will never replace conventional weather stations, for the data from these two sources are essentially complementary and are best interpreted together. Those who believed that meteorological satellites had heralded the coming of the millennium in atmospheric studies were over-optimistic; but these satellites have proved, and will continue to prove, highly important tools in studying both weather and climate.

VII References

Adem, J. 1964a: On the physical basis for the numerical prediction of monthly and seasonal temperatures in the troposphere–ocean–continent system. *Monthly Weather Review* 92, 91–104.

1964b: On the normal state of the troposphere–ocean–continent system in the northern hemisphere. *Geofisica Internacionale* 4(1), 3–16.

1967: Relations among winds, temperature, pressure and density, with particular reference to monthly averages. *Monthly Weather Review* 93, 531–9.

ARACON, 1967: *NIMBUS II user's guide*. Greenbelt, Maryland: NASA. (299 pp.)

Barrett, E. C. 1967a: Weather satellite cloud photography of the British region. *Weather* 22(4), 151–9.

1967b: *Viewing weather from Space*. London: Longmans; New York: Praegers. (154 pp.)

1968a: Notes on the evolution and interpretation of satellite global-scale mosaics. *Weather* 23(5), 198–205.

1968b: *The contribution of meteorological satellites to dynamic climatology*. University of Bristol: unpublished PhD thesis. (206 pp.)

1970a: The estimation of monthly rainfall from satellite data. *Monthly Weather Review* 98, 322–7.

1970b: A contribution to the dynamic climatology of the equatorial eastern Pacific and central America, based on meteorological satellite data. *Transactions of the Institute of British Geographers* 50, 25–53.

Berry, F. A., Bollay, E. and **Beers, N. R.** (editors) 1945: *Handbook of Meteorology*. New York and London: McGraw-Hill. (1,068pp.)

Bjerknes, J. 1969: Atmospheric teleconnections from the equatorial Pacific. *Monthly Weather Review* 97, 163–72.

Booth, A. L. and **Taylor, V. R.** 1968: Meso-scale archive and products of digitized video data from ESSA satellites. *Technical Memorandum* NESCTM-9, National Environmental Satellite Center, Washington, D.C. (20 pp.)

Boucher, R. J. and **Newcomb, R. J.** 1962: Synoptic interpretations of some TIROS vortex patterns: a preliminary cyclone model. *Journal of Applied Meteorology* 1, 127–35.

Clapp, P. F. 1964: Global cloud cover for seasons using TIROS neph-analyses. *Monthly Weather Review* 92, 495–507.

1965: *Parameterization of certain atmospheric heat sources and sinks for use in a numerical model for monthly and seasonal forecasting*. United States Weather Bureau: Extended Forecast Division. (55 pp.) (Mimeo.)

Conover, J. H. 1962: Cloud interpretation from satellite altitudes. *Air Force Cambridge Research Laboratories, Research Note* 81. (77 pp.)

1963: Cloud interpretation from satellite altitudes. *Air Force Cambridge Research Laboratories, Research Note* 81, Supplement 1. (18 pp.)

1964: The identification and significance of orographically-induced clouds observed by TIROS satellites. *Journal of Applied Meteorology* 3, 226–34.

Crowe, P. R. 1949: The trade-wind circulation of the world. *Transactions of the Institute of British Geographers* 15, 37–56.

1950: The seasonal variation in the strength of the trades. *Transactions of the Institute of British Geographers* 16, 24–47.

1951: Wind and weather in the equatorial zone. *Transactions of the Institute of British Geographers* 17, 21–76.

Ericksson, C. O. 1964: Satellite photographs of convective clouds and their relation to vertical wind shear. *Monthly Weather Review* 92, 283–296.

Faller, A. J. 1965: Large eddies in the atmospheric boundary layer and their probable role in the formation of cloud rows. *Journal of Atmospheric Science* 22(2), 176–84.

Fett, R. W. 1964: Aspects of hurricane structure: new model considerations suggested by TIROS and Project Mercury observations. *Monthly Weather Review* 92, 43–60.

Fritz, S. 1965: The significance of mountain lee-waves as seen from satellite photographs. *Journal of Applied Meteorology* 4, 31–7.

Gaby, D. C. 1967: Cumulus cloud lines versus surface winds in equatorial latitudes. *Monthly Weather Review* 95, 203–8.

Gentry, R. C. 1964: *National Hurricane Research Laboratory Report* 69.

Godshall, F. A. 1968: Intertropical convergence zone and mean cloud amount in the tropical Pacific Ocean. *Monthly Weather Review* 96, 172–5.

Hadfield, R. G. 1964: Atlas of cloud vortex patterns observed in satellite photographs. *Stanford Research Institute, Final Report, Contract no.* Cwb-10627. (245 pp.)

Hopkins, M. M. 1967: An approach to the classification of meteorological satellite data. *Journal of Applied Metereology* 6, 164–78.

Hu, M. J-C. 1963: A trainable weather-forecasting system. *Stanford Electronics Laboratories, Technical Report* 6759–1. (44 pp.)

Hubert, L. F. 1963: Middle latitudes of the northern hemisphere, TIROS data as an analysis aid. In Wexler, H. and Caskey, J. E., editor, *Rocket and Satellite Meteorology*, North Holland, 312–16.

Hubert, L. F. and **Krueger, A. E.** 1962: Satellite pictures of meso-scale eddies. *Monthly Weather Review* 90, 457–63.

Johnson, H. M. 1966: Motions in the upper troposphere as revealed by satellite-observed cirrus formations. *National Environmental Satellite Centre Report* 39. (92 pp.)

Jones, J. B. and **Mace, L. M.** 1963: TIROS meteorological satellite operational assists. *Weatherwise* 15, 97–104.

Kendrew, W. G. 1963: *The climates of the continents*. Oxford University Press. (608 pp.) (Fifth edition.)

Kornfield, J., Hasler, A. F., Hanson, K. J. and **Suomi, V. E.** 1967: Photographic cloud climatology from ESSA III and V computer-produced mosaics. *Bulletin of the American Meteorological Society* 48(12), 878–83.

Kuettner, F. P. 1959: The band structure of the atmosphere. *Tellus* 11(3), 267–94.

Landsberg, H. 1945: Climatology. In Berry, F. A., Bollay, E. and Beers, N. R., editors, 1945, 927–98.

Lehr, P. E. 1962: Methods of archiving, retrieving and utilizing data acquired by TIROS meteorological satellites. *Bulletin of the American Meteorological Society* 43(10), 539–48.

Lyons, W. A. and **Fujita, T.** 1968: Mesoscale motions in oceanic stratus as revealed by satellite data. *Monthly Weather Review* 96, 304–14.

Malkus, J., Ronne, C. and **Chaffee, M.** 1960: *Woods Hole Oceanographic Institute, Technical Report* (8 pp.)

Miller, A. A. 1931: *Climatology*. London: Methuen. (318 pp.) (ninth edition.)

Miller, B. I. 1967: Characteristics of hurricanes. *Science* 157, 1389–95.

Namias, J. 1960: Influence of abnormal surface heat sources and sinks on atmospheric behaviour. *Proceedings of the International Symposium on National Weather Prediction, Tokyo, 1960*, 42–65.

Ondrejka, R. J. and **Conover, J. H.** 1966: Notes on the stereo interpretation of NIMBUS II APT photography. *Monthly Weather Review* 94, 611–14.

Palmer, C. E. 1951: Tropical Meteorology. In *Compendium of meteorology*, Boston: American Meteorological Society, 859–80.

Rao, P. K. 1966: A study of the onset of the monsoon over India during 1962 using TIROS IV radiation data. *Indian Journal of Meteorology and Geophysics* 17(3), 347–52.

Riehl, H. 1954: *Tropical meteorology*. New York and London: McGraw-Hill. (392 pp.)

Rosen, C. A. 1967: Pattern classification by adaptive machines. *Science* 156, 38–44.

Sherr, P. E. and **Rogers, C. W. C.** 1965: The identification and interpretation of cloud vortices using TIROS infra-red observations. *ARACON Geophysics Company, Final Report, Contract no.* Cwb-18812. (74 pp.)

Stein, K. 1967: TIROS M design broadens capabilities. *Aviation Week and Space Technology* 22, 5–8.

Sutton, O. G. 1965: The resurgence of interest in the observational sciences. *Weather* 20(6), 174–82.

Thornthwaite, C. W. 1948: An approach towards a rational classification of climate. *Geographical Review* 38, 55–94.

Timchalk, A. 1965: Wind speeds from TIROS pictures of storms in the tropics. *United States Department of Commerce, Meteorological Satellite Report* 33. (34 pp.)

United States Weather Bureau 1965: *Project Stormfury*. Fact sheet. (4 pp.)

Vetlov, I. P. 1966: Role of satellites in meteorology. In *Interpretation and use of meteorological satellite data, WMO Regional Training Seminar, Moscow*. (20 pp.)

Whitney, L. F. 1966: On locating jet streams from TIROS photographs. *Monthly Weather Review* 94, 127–38.

Widger, W. K. 1964: Practical interpretation of meteorological satellite data. *ARACON Geophysical Company, Final Report, Contract no.* AF 19(628)–2471. (385 pp.)

1966: *Meteorological satellites*. New York and London: Holt, Rinehart and Winston. (280 pp.)

Wilcock, A. A. 1968: Köppen after fifty years. *Annals of the Association of American Geographers* 58(1), 12–28.
World Meteorological Office 1966: The use of satellite pictures in weather analysis and forecasting. *WMO Technical Note* 75. (150 pp.)

Prediction of beach changes

by W. Harrison

Contents

I Introduction[1]

1 A rationale for beach prediction

IF valid predictions of changes in beach geometry are ever to become a reality, methods will have to be found for dealing with the complexly interlocked variables that characterise the beach–ocean–atmosphere system. Extrapolation to natural beaches of formulae obtained in laboratory experiments has to date been a failure; so likewise have attempts to predict beach changes using theoretical formulae. The laboratory derived relationships between wave action and beach profile characteristics obtained, for example, in such excellent studies as those of Bagnold (1940) and Kemp (1961) depend upon the attainment of steady-state conditions, and thus have limited predictive value for the short-memory system of a natural beach. In addition to the usual problems of scale effects between field and model, the modelling of the groundwater table is either ignored or inadequate in most laboratory studies, and in all theoretical treatments to date.

An approach under development in the United States takes nature as the laboratory. The beach environment is interrogated for a large number of variables of interest, and over a sufficiently long period to permit computerised time-series or other correlation-analysis techniques to be applied to the data gathered. In a sense this approach may be thought of as one in which the data are in search of a model. Early work on this approach has been done by Harrison and Krumbein (1964), Harrison and Pore (1964), Dolan (1965) and Harrison, Pore and Tuck (1965).

Although physical models may indeed be inferred from such an approach, the results immediately hoped for will be empirical equations that will permit the prediction of variations in the geometrical aspects of concern, for the specific beach in question. The physical meaning of the coefficients in the fitted equations may not be readily apparent, nor may they in fact enter linearly into the true physical relationship. Nevertheless the linear regression methods employed in the approach permit identification of the variables that play strong roles in the system investigated. And the equations developed for the interrelated variables, although based only on the linear model, provide an initial step towards successful prediction.

2 Approach

A time series of observations of beach processes and responses is obtained from a beach of interest. The observations include measurements of as

[1] Contribution 341 of the Virginia Institute of Marine Science, Gloucester Point, Virginia 23062.

many physically meaningful variables as possible, and the values are then used in multivariate statistical analysis in a search for significant relationships between the 'process' and 'response' variables. The results give insight into the *relative* importance of the several 'independent' variables measured, for the specific beach conditions in effect during the study period. The insights obtained from one beach may be applied also in the design of studies on other beaches in similar environments. And predictor equations for process–response relationships on the beach in question will be obtained, which may form a basis for 'forecasting' specific beach responses when process conditions fall in the same range as those that existed when the basic time-series of measurements was obtained.

II Example: prediction of geometrical changes on a two-dimensional ocean beach over a tidal-cycle interval

1 Basic time-series

The 26-day-long time-series of observations that will be used in this section has been published elsewhere (Harrison, *et al.*, 1968). It is for an Atlantic Ocean beach at Virginia Beach, Virginia (Figure 1). The following features are notable about this time-series:

1 no measurable rainfall was recorded during the entire 26-day period
2 generally light winds of variable direction prevailed throughout the period of study
3 wide ranges in values of wave height, period, crest length and steepness were noted during the study period
4 the foreshore was generally plane throughout the study period; only one set of broad cusps formed
5 no progressive sand waves were observed moving along the shoreline (cf. Harrison *et al.*, 1968, plates A–I)

Thus the effects of rainfall on fluctuations in the water table in the foreshore can be discounted and the effect of winds on the nearshore water circulation minimised. This reduces the complexity of the problem in that the wind and rainfall variables can be left out of the analytical work. Examples of cases in which wind variables were used in an analysis may be found in Harrison, Pore and Tuck (1965).

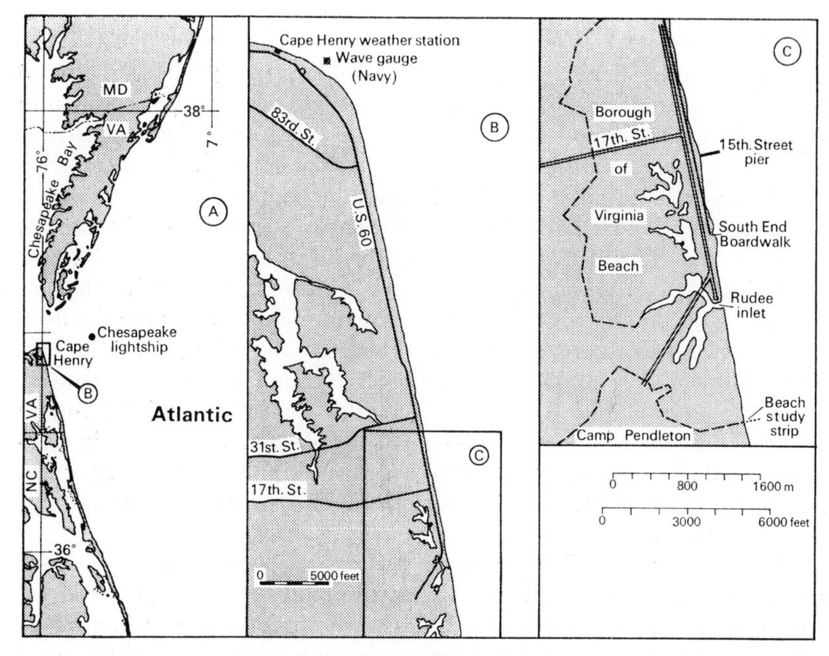

Fig. 1 Maps showing location of area of investigation and beach study strip (map **C**) where measurements were made.

Additional information on how the variables were defined and the details of the measurement techniques may be found in Harrison *et al.* (1968).

2 Dependencies investigated

The dependency of each of the following 'responses' in beach geometry on several dimensionless ratios formed from beach 'process' variables, will be investigated:

1. change in quantity of foreshore sand during a tidal-cycle interval (as determined from the *geometry* of successive foreshore profiles)
2. the advance or retreat of the shoreline (mean high water line) during a tidal-cycle interval
3. the slope of the foreshore, as measured at the end of a tidal-cycle interval.

Because of the lag in time between the inception of a given process and its effect on the foreshore geometry, as measured at the end of the tidal cycle, provision will be made in the analytical procedure for lagging the process variables through time, where necessary.

3 Change in quantity of foreshore sand, ΔQ_f

As suggested by field measurements (Strahler, 1964; Giese, 1966) on beaches near equilibrium, erosion in the mid-swash zone and accretion in the regions of the step and the limit of the swash probably represent a continuous process on the foreshore of steep tidal beaches. The mean foreshore slope must attain a certain minimum inclination before this process is clearly developed, and although the angle is not definitely known, the process is difficult to observe where the mean foreshore slope is lower than about 4°–5°. When beaches are in equilibrium, progressive changes in foreshore geometry are accomplished by transfer of sandy material from the mid-swash to the reach and to the step; as the tidal plane rises and falls, the geometry of the foreshore goes through a series of changes, returning to the initial configuration after an interval of tidal-cycle length. Duncan (1964) has documented the short-term changes in foreshore profile on a gently sloping tidal beach and emphasised the important role of the beach water table characteristics on erosion and deposition in the swash zone.

The morphological feature known as 'the step' is seldom observed on the relatively gently sloping beach at the study strip (Figure 1C) on Virginia Beach. The processes of mid-swash erosion and step migration are likewise rarely seen. It has been necessary to resort to detailed beach profiles, repeated at four-hour intervals, to document the general trends in morphological changes—i.e., changes resulting from either net erosion or deposition, or from the simple shifting of sand on the foreshore, over a tidal-cycle interval, without net erosion or deposition (equilibrium beach).

An equilibrium beach is a 'two-dimensional' beach in which as much sediment is supplied by beach drifting as is lost by this process. It also embodies a 'closed' swash–breaker zone sediment system in which there is no net transfer of sediment across either boundary. Thus an instantaneous, non-equilibrium foreshore (one which is being eroded or built up) is one for which there is a net transfer of sediment across the breaker zone. In this study a two-dimensional beach is a necessary assumption.

Given sufficient resolution of changes in the altitude of the foreshore at reference points, the amount of beach erosion or deposition has been determined (cf. Duncan, 1964; Strahler, 1964) over relatively short periods of every half hour, or over an interval of half-tidal-cycle length (Inman and Rusnak, 1956, 26). In this study the net change in quantity of foreshore sand (beach erosion or deposition) is determined for the dynamic unit of one tidal cycle (for example, low water to high water to low water), and the processes effecting a change in foreshore geometry are monitored at intervals throughout the tidal cycle. The region of the foreshore studied lies between the inshoremost breaker zone of successive low

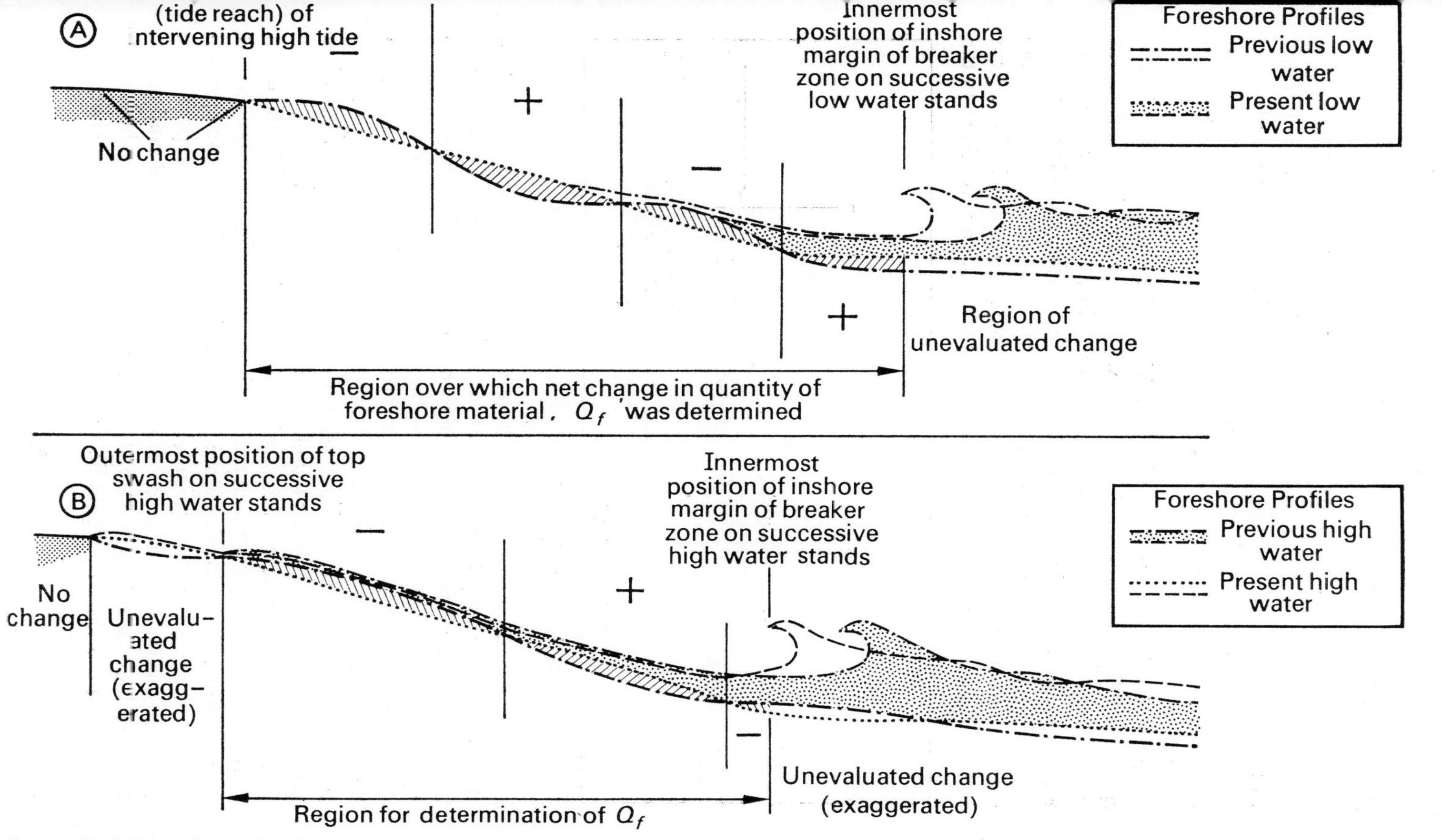

Fig. 2 Definition sketch for determination of ΔQ_f: **A,** for low-water-to-low-water interval; **B,** for high-water-to-high-water interval.

water stands (Figure 2A) and the limit of the swash of the intervening high water stand. In the case of the tidal-cycle interval from one high tide to the next, the region of the foreshore studied lies between the offshore-most position of the limit of the swash, at successive high water stands, and the inshoremost position of the breaker zone on the successive high tides (Figure 2B).

In considering the measurable quantities upon which changes in the quantity of foreshore sand ΔQ_f depend, we may begin with the forces at work. Neglecting the effects of wind, these will be primarily the forces related to the impinging waves. Thus breaker height, H_b (Figure 3) and breaker period, T_b, will be of prime importance as they determine the amount of water translated across the breaker zone per unit time. Next in importance is the angle of approach of the breaking wave front to the

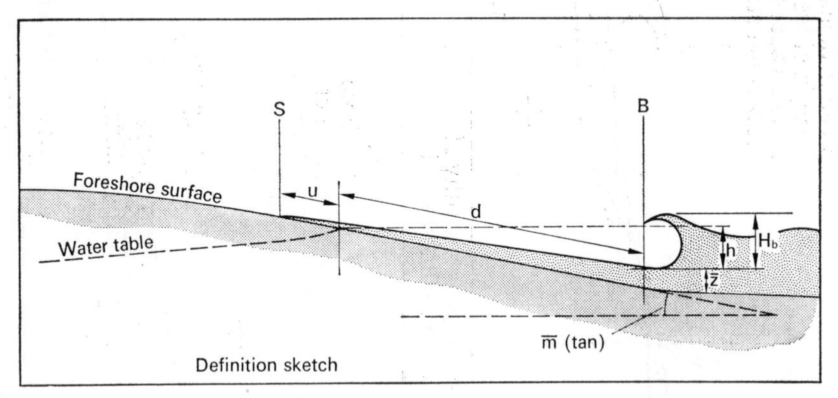

Fig. 3 Definition sketch for variables measured at beach study strip (Fig. 1C).

shoreline, because the amount of runup of the swash depends in part on this angle. This may be expressed as a ratio r_1 of two lengths. We can approximate the form of the breaking wave (Figure 3) in terms of a ratio r_2 of the height of the breaker to the trough-to-bottom distance, z, in front of the breaking wave, H_b/z. As an example, a plunging wave will have a large H_b/z ratio. Finally we must consider the duration of application of the wave energy upon any one zone of the sloping foreshore. This is a function of the rate of rise or fall of the still water level $\pm\eta$.

The essential geometry of the central foreshore is bound up in the mean slope between the swash reach and the inshore margin of the breakers at any one time. The mean foreshore slope ($\bar{m}$, Figure 3) is the ratio r_3 of two lengths. Among the most essential features of the grains composing the foreshore are measures of mean grain diameter $\bar{D}$ of surficial sand on the mid-foreshore, and the grain (solid) density, ρ_s, of this mid-foreshore sand. Although fluid viscosity would be a better measure, a sufficient characterisation of the mechanical behaviour of the fluid is its density, ρ_l. With regard to the effects of groundwater flow and consequent lift forces at the free surface of the sloping foreshore, we may construct the ratio (Figure 3) of the height of groundwater head, h, to be dissipated over the distance d between the outcrop of the water table and the breakers. Thus, r_4 is h/d. A final ratio r_5 takes into account the relative percentage of unsaturated foreshore $\bar{u}$ (Figure 3) to saturated foreshore d, and is formulated to account for the effects of percolation of the swash on grain deposition near the swash limit. Information relative to all of the foregoing variables is given in Table 1.

If we have not overlooked any important feature, there must be some definite quantitative relation connecting the foreshore volume change ΔQ_f with the other quantities enumerated, and it may be symbolised by the equation:

$$F(\Delta Q_f, H_b, T_b, \pm\eta, \bar{D}, \rho_s, \rho_l, r_1 \ldots r_5) = 0 \qquad (1)$$

If equation (1) is to be dimensionally homogeneous, it must contain identical powers of each of the respective dimensions. Therefore, we express all of the independent variables in the three dimensions of M, L and T and apply the Π-theorem to make it dimensionally homogeneous,

Equation (2) is a dimensionally correct form of equation (1):

$$\Delta Q_f = F\left(\frac{\bar{H}_b}{\pm\eta\bar{T}_b}, \frac{\rho_l}{\rho_s}, \frac{\bar{H}_b}{\bar{D}}, r_1 \ldots r_5\right)\bar{H}_b^{\,3} \qquad (2)$$

in which the bars above H_b, T_b and ρ_s indicate average values, as explained in the section following.

The nature of the dependence expressed by the unknown function F, however, remains to be determined from experimental (field) investigations. Because ΔQ_f is determined by X_i (independent variables) acting over the previous 12·5-hour tidal-cycle interval, function F will be investigated by a linear multiple regression 'screening' procedure, in which each dimensionless Π term is allowed to enter in over eight lag periods each separated by 1·5 hours. The problem is solved by calculating:

$$\Delta Q_{f_{0\cdot0}} \Big/ H^3_{b_{0\cdot0}}, \ \Delta Q_{f_{0\cdot0}} \Big/ H^3_{b_{1\cdot5}}, \ldots, \Delta Q_{f_{0\cdot0}} \Big/ H^3_{b_{12\cdot0}}$$

where the subscripts for the variables indicate lag times in hours. Thus in addition to a dependent variable for $t_{0\cdot0}$, there will be 8 dependent variables for lag periods $t_{1\cdot5}, t_{3\cdot0}, t_{4\cdot5}, \ldots, t_{12\cdot0}$. The matrix of independent variables will be composed of the 8 variables within the parentheses of equation (2), each taken over 8 lag times ($t_{1\cdot5}, \ldots, t_{12\cdot0}$). Thus for each Y_i, 64 independent variables will be screened. Predictor equations containing the strongest combinations of X_i, for X from 1 to 6, will be ranked

Table 1 Variables used in this study that were measured or derived in development of the basic time-series, their dimensions, schedule of measurements and ranges in value

Symbol	Dimensions	Description	Measurement schedule	Range in values	Probable error
b	L	Breaker distance, or 'runup'. [Length (Figure 4) along foreshore surface from inshore margin of breakers to top of swash]	Derived (every 4 hours)	7·5 to 61·7 m	±1·5 m
d	L	Distance (Figure 3) between outcrop of water table on foreshore and inshore margin of breakers	Derived (every 4 hours)	6·0 to 57·0 m	±1·5 m
$\bar{D}$	L	Mean nominal grain diameter of sand samples taken at mid-foreshore	Every 6·25 hours (high and low water)	0·187 to 0·411 mm	±0·02 mm
h	L	Hydraulic head; vertical distance (Figure 3) between horizontal plane passing through water table outcrop and horizontal plane through bottom of trough in front of breaking wave	Derived (every 4 hours	−0·20 to +1·60 m	±0·04 m
$\bar{H}_b$	L	Mean height (Figure 3) of breaking waves	Every 4 hours	0·15 to 0·89 m	±0·06 m, low waves; ±0·12 m, high waves
H_{bs}	L	Height of significant breaking waves	Derived from H_b (every 4 hours)	0·20 to 1·20 m	±0·06 m, low waves; ±0·12 m, high waves
$\bar{m}$	O	Mean slope (Figure 3) of foreshore (tan)	Derived from elevation data	0.0100 to 0·1263 (0·59° to 7·2°)	±0·2°

t_w	O	Temperature of sea water	Every 4 hours	21·1 to 26·4°C	±0·2°C
$\bar{T}_b$	T	Mean period of breaking waves	Every 4 hours	4·0 to 12·4 seconds	±0·2 seconds
u	L	Unsaturated beach surface (Figure 3) between outcrop of water table on foreshore and swash reach	Derived (every 4 hours)	0·0 to 11·8 m	±0·3 to ±1·0 m for low and high values, respectively
$\bar{z}$	L	Mean trough to bottom distance (Figure 3) in front of a breaking wave	Every 4 hours	0·07 to 0·98 m	±0·005 m
$\bar{\alpha}_b$	O	Mean acute angle (tan) between shoreline and crest of breaking wave	Every 4 hours	0·0000 to 0·3172 (0·0° to 17·6°)	±1·0 to 2·0°, for a single wave
ΔQ_f	L^3	Quantity of sand (Figure 2) eroded from or deposited on foreshore in one tidal cycle	Derived every 6·25 hour (highs and low water)	$-5·73$ m³ to $+3·00$ m³	±0·05 m³
ΔS	L	Change in shoreline (mean high water line) (+ = advance; − = retreat)	Derived (every low water)	−3·1 to +2·2 m	±0·075 m
$\pm\eta$	O	Rate of rise (+) and fall (−) of still water level at tide gauge	Derived from tide-curve data	−0·42 to +0·39 m/hour	±0·015 m/hour
ρ_l	ML^{-3}	Density of liquid (sea water); derived from temp. and salinity data	Every 4 hours	1·0169 to 1·0202 g/cm³	±0·00005 g/cm³
ρ_s	ML^{-3}	Density of solids (sand grains) on mid-foreshore	Every 6·25 hours (high and low water)	2·6172 to 2·6740 g/cm³	±0·00005 g/cm³

Source: Harrison *et al.* (1968)

according to the total reduction of variance (R^2) criterion, and the predictor equation that best explains the dependence of Y_i on the X_i will be assumed to exemplify the significant dependent relationships under the conditions of the investigation.

The procedure adopted for selecting predictors involves expressing ΔQ_f as a linear function of a number of variables (predictors X_n ($n = 1, \ldots, N$).

Thus:

$$\Delta Q_f = A_0 + A_1X_1 + A_2X_2 + \ldots A_nX_n + \ldots A_NX_N$$

where the coefficients A_n ($n = 0, \ldots, N$) are determined by the method of least squares. (In the scheme of multiple regression analysis described below, the significance of improvement attained at each step of the analysis is not tested by F ratios for reasons given in Harrison, Pore and Tuck [1965, 6106]. Also, a rule of thumb [cf. Harrison and Pore, 1967, Figure 19] suggests that the number of cases, N, analysed should be about seven times the number of X_i chosen.)

The 'screening procedure' (cf. Miller, 1958) is used for the multiregression analysis of this study. Basically the technique is shown below:

$$\Delta Q_f = A_1 + B_1X_1 \quad (3)$$
$$\Delta Q_f = A_2 + B_2X_1 + C_1X_2 \quad (4)$$
$$\Delta Q_f = A_n + B_nX_1 + C_{n-1}X_2, \ldots, NX_n \quad (5)$$

where the As are constants, and B_1, B_2, C_1, C_2, etc., are regression coefficients.

The procedure is first to select the best single predictor (X_1) for regression equation (3). The second regression equation (4) contains the X_2 that contributes most to reducing the residual after X_1 is considered. This is usually, but not always, the best subset of X_i out of the original set. In the somewhat analogous field of meteorology, studies have shown that by this screening procedure a highly reliable set of predictors can be selected.

One limitation of the linear analysis is that some variables that have only a small linear effect may become quite strong in a model that explicitly includes non-linear effects. It is also true, however, that the linear model is generally the best one for initial work with many variables.

In employing multiregression techniques on the expression:

$$\Delta Q_f/(\bar{H}_b^3)_{0-8} = f\left[\left(\frac{\bar{H}_b}{\eta \bar{T}_b}\right)_{0-8}, \left(\frac{\rho_l}{\rho_s}\right)_{0-8}, \left(\frac{\bar{H}_b}{\bar{D}}\right)_{0-8}, (\tan \bar{m})_{0-8}, (\tan \bar{\alpha}_b)_{0-8}, \left(\frac{\bar{H}_b}{\bar{z}}\right)_{0-8}, (u/d)_{0-8}\ (h/d)_{0-8}\right] \quad (6)$$

it is important to recognise that the correlations between $\Delta Q_f/\bar{H}_b^3$ and the ratios $\bar{H}_b/\eta\bar{T}_b$, $\bar{H}_b/\bar{D}$, and $\bar{H}_b/\bar{z}$ may be spurious (cf. Benson, 1965), because of the commonality of the variable $\bar{H}_b$ in the ratios. Table 2 summarises the results for selection of the predictors in the screening run that gave the strongest four-predictor equation.

Table 2 Selection of predictors (for $Y = \Delta Q_f/H_b{}^3$) by screening process (lag interval expressed in hours)

Equation	Variable	Low water to low water ($N = 48$) Sign of correlation	Lag (hours)	R^2
(1)	$\rho_l/\rho_s)$	+	6·0	0·15
(2)	$\bar{\alpha}_b$	−	4·5	0·28
(3)	h/d	−	10·5	0·38
(4)	u/d	+	3·0	0·49

The actual predictor equation for this run is:

$$\Delta Q_f = [-4293{\cdot}186 + 11367{\cdot}022(\rho_l/\rho_s)_{-6\cdot0} - 257{\cdot}425(\tan\bar{\alpha}_b)_{-4\cdot5} - 2292{\cdot}206(h/d)_{-10\cdot5} + 185{\cdot}286(u/d)_{-3\cdot0}](\bar{H}_b{}^3)_{-9\cdot0} \qquad (7)$$

where the lag times are in hours before low water and ΔQ_f is in m³.

It is of interest that the strongest (first chosen) predictor for the low water interval (Table 2) is the ratio ρ_l/ρ_s, manifesting at time of high water (lag 6·0 hours), and exhibiting a positive correlation with ΔQ_f. Thus, when the liquid density is high relative to particle density, during time of high water, more sand grains will be carried above the outcrop of the groundwater table to be deposited in the region of the swash reach, where the beach will be built up ($+\Delta Q_f$), and *vice versa*. Variations in ρ_l/ρ_s may be due to the variation in temperature and salinity of the sea water, as modulated by tidal flow in and out of Chesapeake Bay, to variations in particle size, or to both factors. (The finer sand sizes contain heavy minerals, while the coarser grains contain shell fragments which are of lesser density than the quartz sand that makes up the great bulk of the foreshore material.) The lack of appearance of ρ_l/ρ_s in other lag periods indicates its limited importance to ΔQ_f and its restriction to the high water lag. Thus, although it is realised that the 'dimensional analysis' format is not a very desirable one for the multiregression screening procedure, it has not been entirely useless here. In the present exercise, for example, we see that changes in the quantity of foreshore sand over the interval of a tidal cycle, while depending rather strongly upon $\bar{H}_b$ at time of rising half tide (nine hours before the end of the low tide interval, equation (7)), are significantly correlated also with the ratios ρ_l/ρ_s, $\tan\alpha_b$, h/d, and u/d: that is, ΔQ_f depends upon the energy term of breaker height, upon the geometry of the application of that energy ($\bar{\alpha}_b$) shortly after the beginning of falling tide, upon the difference in the density between the material to be transported and the transporting medium (ρ_l/ρ_s), upon the hydraulic head acting through the foreshore surface (h/d), and upon the length (if any) of unsaturated sand traversed by the swash at the time of falling half tide. Thus for the 12·5-hour time scale of the tidal-cycle interval, the regression analysis suggests that these variables, above all others set forth in (2), are the most important for determining changes in the quantity of foreshore sand in a beach environment like the one studied.

But what might we find if, instead of investigating a dimensionally homogeneous statement like equation (2), we simply screen a series of variables which, although not part of a dimensionally homogeneous *equation*, are nevertheless themselves dimensionless and—at the same time —have the important feature of a complete lack of commonality between the dependent and any of the independent variables? Further, each of these dimensionless variables may be cast as a ratio that we feel to be of physical significance in its own right.

Thus instead of the dimensionless variable $\bar{H}_b/\eta\bar{T}_b$ of equation (2), we might utilise $(\bar{H}_b/g\bar{T}_b^2)^{\frac{1}{2}}$. The latter variable (Galvin, 1968) is a reasonable expression of average breaker 'steepness', and is obtained by dividing the mean breaker height by a hypothetical mean wave length at breaking. This mean wave length is obtained by multiplying the equivalent solitary wave speed by the mean wave period. The resulting quantity can be reduced to $(\bar{H}_b/g\bar{T}_b^2)^{\frac{1}{2}}$, omitting the dimensionless factors describing breaker depth-to-height ratio and the depression of the trough below the mean water level.

Galvin (1968) used $(H_b/gT_b^2)^{\frac{1}{2}}$ as an index of 'breaker steepness' in an attempt to develop a classification of breaker types on laboratory beaches. He found that breaker type goes from spilling to plunging to 'surging' as the breaker steepness decreases or the beach slope increases. He also found that the best parameter for breaker type classification included the beach slope $(H_b/gmT_b^2)^{\frac{1}{2}}$, but because we wish to make a separate assessment of the effect of beach slope on ΔQ_f, the mean beach slope will be inserted as a separate variable.

Another dimensionless variable that would be expected to have physical significance is the hydraulic head divided by the breaker distance, h/b. As shown in Figure 4, this ratio becomes smaller the farther the swash reach moves up the foreshore surface, for a constant value of b. Breaker distance, or 'runup', will have quite different effects on the movement of beach sand on the foreshore, depending upon whether the runup traverses saturated or unsaturated foreshore sands. And the magnitude of the groundwater head at a given instant may have considerable influence on the ease with which sand grains may be eroded from the foreshore surface. For a constant foreshore slope and hydraulic head, breaker distance is primarily a measure of the energy of the swash available to transport sand up or down the foreshore. The interplay of breaker distance and hydraulic head on the erosion and deposition of foreshore sand is complex, and the ratio h/b is but a crude estimate of the effects.

Besides b, another characteristic of the swash that should enter into our analysis is $\bar{z}$, or the mean trough to bottom distance in front of the breaking

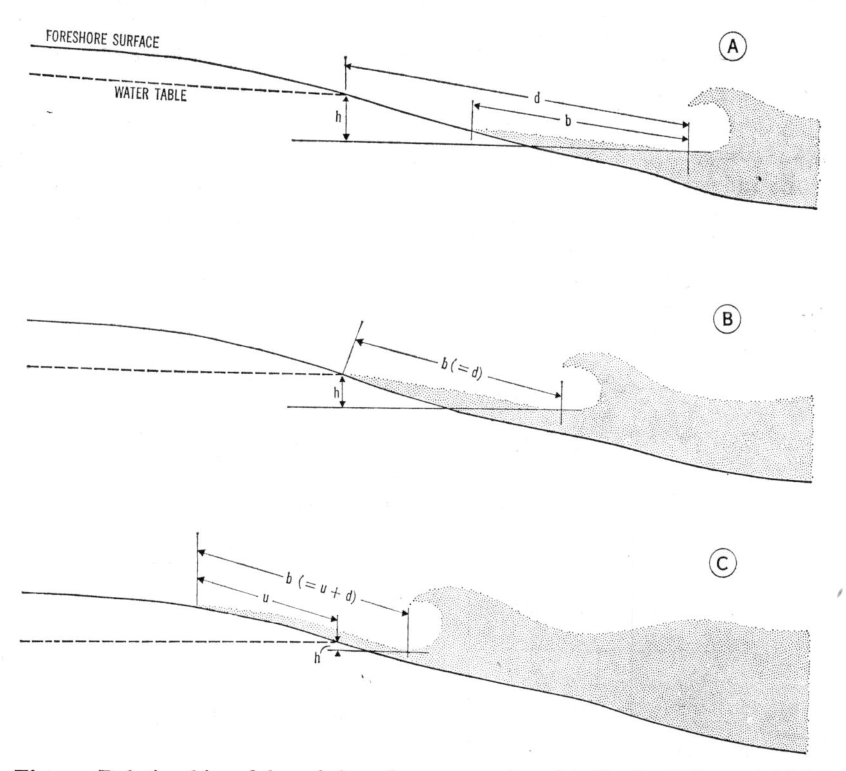

Fig. 4 Relationship of h and b at low water **A,** mid-tide level **B,** and high water **C.**

wave. This value is a measure of the mean swash depth (for a constant slope) and, as an indication of the thickness of the swash, reflects the prevailing mean shear stresses available in the swash zone. This variable may be combined with a grain characteristic $\bar{D}$, or the mean grain diameter of sand from the surface of the mid-foreshore. The value of $\bar{D}$ enters into consideration of roughness of the foreshore, the sizes of grains in suspended transport, and the rate of groundwater flow through the foreshore or percolation of swash water into it. Thus the ratio $\bar{D}/\bar{z}$ is a dimensionless parameter that reflects the interplay of grain size and swash characteristics as these factors relate to changes in the quantity of foreshore sand.

An additional characteristic of both grain and the fluid medium is the ratio ρ_l/ρ_s, found to be significant earlier. Finally, we include $\bar{\alpha}_b$ in our relationship, as before. Thus the relationship to be tested now for five times, three hours apart, for each variable is:

$$\Delta Q_f = f[\bar{H}_b/g\bar{T}_b^{\,2})^{\frac{1}{2}}{}_{0-4}, (\rho_l/\rho_s)_{0-4}, (h/b)_{0-4}, (\bar{D}/\bar{z})_{0-4}, (\bar{m})_{0-4}, (\bar{\alpha})_{0-4}] \quad (8)$$

Only the results of the screening run for the interval ending on low tide are presented in Table 3. The R^2 value for the first *six* predictors for the 12·5-hour interval ending on *high* tide is only 0·40, indicating that the six-predictor equation explains too small a percentage of the total variance to be judged significant. The first six variables selected for the high tide interval were $(\bar{H}_b/g\bar{T}_b^{\,2})^{\frac{1}{2}}{}_{0\cdot 0}$, $(\bar{m})_{0\cdot 0}$, $(\rho_l\rho_s)_{12\cdot 0}$, $(\bar{\alpha}_b)_{12\cdot 0}$, $(h/b)_{12\cdot 0}$, and $(\bar{D}/\bar{z})_{3\cdot 0}$, in that order.

The five-predictor equation obtained by screening eight, for the low water interval, is:

$$\Delta Q_f = 5{\cdot}803 - 113{\cdot}151(\bar{H}_b/g\bar{T}_b^{\,2})^{\frac{1}{2}}{}_{0\cdot 0} - 52{\cdot}111(h/b)_{3\cdot 0} - 7{\cdot}269(\bar{\alpha}_b)_{6\cdot 0} + 2{\cdot}724(\bar{D}/\bar{z})_{6\cdot 0} - 35{\cdot}067(\bar{m})_{12\cdot 0} \quad (9)$$

Table 3 Selection of predictors (for $Y = \Delta Q_f$) by screening procedure (lag interval expressed in hours before low water)

Equation	Variable	Low water to low water ($N = 48$)		
		Sign of correlation	Lag (hours)	R^2
(1)	$(\bar{H}_b/g\bar{T}_b^{\,2})^{\frac{1}{2}}$	−	0·0	0·18
(2)	h/b	−	3·0	0·34
(3)	$\bar{\alpha}_b$	−	6·0	0·44
(4)	$\bar{D}/\bar{z}$	+	6·0	0·50
(5)	$\bar{m}$	−	12·0	0·55
(6)	$(\bar{H}_b/g\bar{T}_b^{\,2})^{\frac{1}{2}}$	−	3·0	0·58
(7)	$(\bar{H}_b/g\bar{T}_b^{\,2})^{\frac{1}{2}}$	+	6·0	0·61
(8)	h/b	−	6·0	0·64

The fact that ρ_l/ρ_s, the strongest predictor in (7), is not selected as one ofthe first eight predictors in (9) is related to the fact that both $(\bar{H}_b/g\bar{T}_b^{\,2})^{\frac{1}{2}}$ and h/b are stronger (in terms of R^2) predictors of the dependent variable than is ρ_l/ρ_s (cf. Table 2). And the selection of subsequent predictors (X_i) by the screening process is conditioned by their *joint* contribution to reduction of variance of ΔQ_f, along with the first two variables chosen.

Thus, while in terms of expression (7) ρ_l/ρ_s is the most important ratio, in terms of (9) it is not.

The physical reasonableness of the predictor–predictand correlations within equation (9) will be examined now. Breaker steepness, at the end of the tidal-cycle interval (low water, lag 0·0 hours, Table 3), is negatively correlated with ΔQ_f. The inverse correlation indicates that as breaker steepness increases, erosion ensues. The time of the correlation indicates that this relationship has its most pronounced effect upon ΔQ_f at the end of the low water to low water interval. This negative correlation exists also at time of falling half tide (Table 3, lag 3·0 hours), but the correlation becomes positive at time of high water (lag 6·0 hours) midway through the tidal-cycle interval. Thus net deposition (positive ΔQ_f) on the foreshore is correlated with *steep* breakers at time of high water. Such a correlation is believed related to the fact that as a steep breaker erodes the mid-foreshore, its uprush crosses the outcrop of the water table with a greater suspended load than that of a low breaker. As the steep breaker's uprush percolates into the unsaturated sand it deposits a greater portion of suspended load than would a low breaker. Thus the beach builds up more under steep breakers at time of high water.

The negative correlation (Table 3) of h/b with ΔQ_f likewise reflects the fact that a relatively large amount of swash water (b) passing over the unsaturated foreshore produces a relatively great amount of deposition ($+\Delta Q_f$). Thus, as breaker distance becomes large relative to h (cf. Figure 4C), sand grains are carried onto the unsaturated foreshore by the uprush. Here they may be deposited when percolation of part or all of the swash water decreases the potential energy of the swash (at time of maximum excursion of the uprush) to the point where the backwash will be unable to carry some of the grains back down the slope. The negative correlation between h/b and ΔQ_f is present (Table 3) at both falling half tide and at high water, but the strength of the correlation is greatest at falling half tide. This is probably in part reflection of the fact that the effect of imbalances in the h/b ratio will leave a more lasting signature upon the foreshore when they manifest during the time of falling half tide (cf. $+u/d$, Table 2), and partly due to the fact that when h/b is relatively large at the time of falling half tide the tendency is towards erosion because the larger pore-water pressure along the free surface of the foreshore augments the downslope mobility of the grains. (This last assessment holds only when the swash is traversing saturated foreshore sands, exclusively.)

The negative correlation between $\bar{\alpha}_b$ and ΔQ_f (Table 3, high water lag) is again explicable in terms of considerations of runup and percolation of swash. Thus, as $\bar{\alpha}_b$ becomes smaller and the breaking wave fronts are more nearly parallel to the shoreline, the quantity of swash water and the distance of its translation up the foreshore increase. As expected, the effect upon ΔQ_f will be most pronounced at time of high water, when the effect of wave runup over unsaturated foreshore sands will also be most pronounced. At the time of high water, therefore, as $\bar{\alpha}_b$ increases, ΔQ_f tends to decrease, and *vice versa*.

The positive correlation between ΔQ_f and $\bar{D}/\bar{z}$, at time of high water

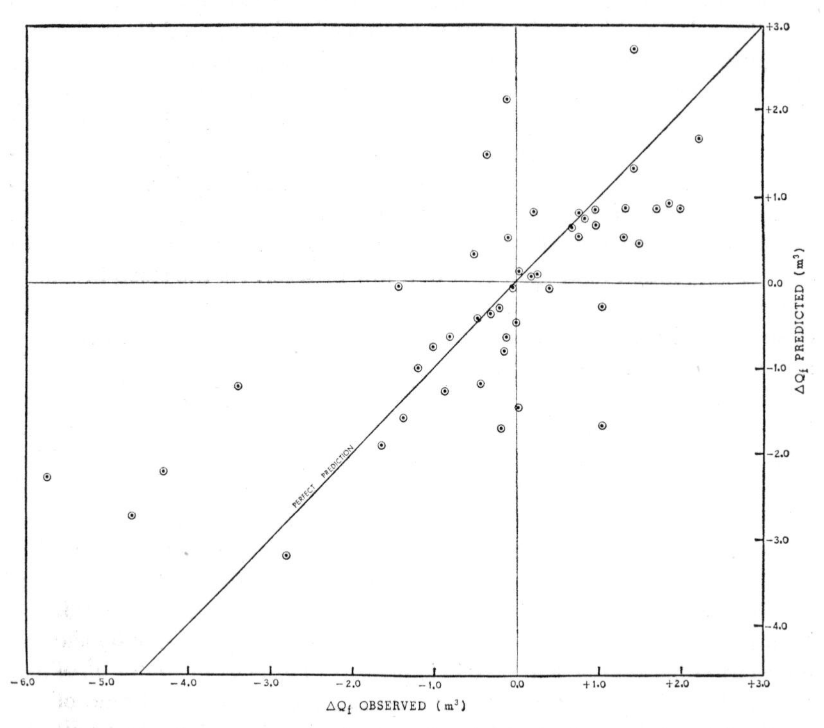

Fig. 5 Predicted versus observed values for change in the quantity of foreshore sand, from one low-water stand to the next, using equation (9).

(−6·0 hours, Table 3), is related to the fact that as the size of the grains on the foreshore becomes large, for a constant swash depth ($\approx \bar{z}$), percolation of swash into the unsaturated foreshore is enhanced and net deposition of sand transported by the uprush is enhanced.

Finally the regression analysis indicates that the foreshore slope exerts its greatest effect upon ΔQ_f at the time of the previous low water. This relationship between $\bar{m}$ and ΔQ_f (Table 3) reflects the control of the initial beach slope over the events to follow, and these events determine the probability for erosion or deposition in the ensuing 12·0 hours. The negative correlation implies that the lower the initial slope, the more deposition ($+\Delta Q_f$) that may be expected, and the steeper the initial slope, the greater the probability for net erosion over the next 12·0 hours. For average wave, sea water, sediment, and groundwater characteristics, this relationship is generally the correct one: for example, an initially steep slope undergoes scarping and net erosion, while a low initial slope leads to heightened grain transport up the foreshore and to net deposition.

In view of the foregoing, expression (9) may be considered a reasonable statement of the dependency of foreshore volume changes on the 'critical' environmental variables, as such variables were measured at Virginia Beach and cast in the formulation of (8). Because of the complicated interdependencies between the X_i and Y, and between the X_i themselves, and owing to the 'feedback' mechanisms in the beach environment, the author feels that the best expression for changes in the mass (geometry) of the foreshore will come from empirical correlations of reliable *field* data, such as the correlations presented in equation (9). A plot of predicted versus observed values for equation (9) for ΔQ_f is given in Figure 5.

4 Advance or retreat of the shoreline (mean high water line), ΔS

Attention is turned now to another geometrical response of the foreshore, the problem of the advance or retreat of the shoreline, ΔS. It may be assumed that the seaward ($+S$) or landward ($-S$) movement of the mean high water line (shoreline) will depend upon the same variables as ΔQ_f. We shall document the motion of the 'shoreline' as it changes during the interval of one tidal cycle, extending from one low water to the next. The magnitude of advance or retreat of the shoreline will be expressed as a function of the same X_i as in (6), excluding only $\bar{m}$. (There is considerable data redundancy between ΔS and $\bar{m}$ and it is unlikely that there will be a change in the position of the shoreline without there being a change in the foreshore slope). Thus for five lag periods, each separated by three hours:

$$\Delta S = f[(\bar{H}_b/g\bar{T}_b{}^2)^{\frac{1}{2}}{}_{0-4}, (\rho_l/\rho_s)_{0-4}, (\tan \bar{\alpha}_b)_{0-4}, (h/b)_{0-4}, (\bar{D}/\bar{z})_{0-4}] \quad (10)$$

Although always defined arbitrarily, the 'shoreline' (cf. Adams, 1942, 693) is most often associated with the mean high water line, as determined during times of average tidal range and beach slope. The 'mean high water line', for the 26-day time-series of this study, was determined to lie 0·91 m above MSL (Harrison *et al.*, 1968, Figure 3). This mean high water line is located within a level plane which lies, for the time-series of observations studied,

1 just below the lowest position of the top of the swash for all high water stands

2 just above the highest position of the inshore margin of the breakers for all high water stands (excepting two)
3 just above the highest position of the top of the swash for all low water stands (excepting one).

It is the onshore or offshore movement of the point (in two dimensions) formed by the intersection of the imaginary level plane with the foreshore surface that is investigated in (10). The results of screening (10) appear in Table 4.

Table 4 Selection of predictors (for $Y = \Delta S$) by screening procedure (lag interval expressed in hours before low water; $N = 48$)

Equation	Variable	Sign of correlation	Lag (hours)	R^2
(1)	$\bar{D}/\bar{z}$	−	0·0	0·19
(2)	$(\bar{H}_b/g\bar{T}_b^2)^{\frac{1}{2}}$	−	3·0	0·27
(3)	$\bar{\alpha}_b$	−	6·0	0·33
(4)	$\bar{\alpha}_b$	+	0·0	0·38
(5)	h/b	−	3·0	0·43
(6)	$\bar{D}/\bar{z}$	+	6·0	0·47

The complete prediction equation for the 6 variables of Table 4 is:

$$\Delta S = 2{\cdot}312 - 0{\cdot}909(\bar{D}/\bar{z})_{0\cdot 0} - 61{\cdot}037(\bar{H}_b/g\bar{T}_b^2)^{\frac{1}{2}}_{3\cdot 0} - 5{\cdot}708(\bar{\alpha}_b)_{6\cdot 0} + 4{\cdot}682(\bar{\alpha}_b)_{0\cdot 0} - 22{\cdot}190(h/b)_{3\cdot 0} + 1{\cdot}874(\bar{D}/\bar{z})_{6\cdot 0} \quad (11)$$

In Table 4, we see that 2 variables appear twice, $\bar{D}/\bar{z}$ and $\bar{\alpha}_b$. They appear at time of high water or at the time of low water (0·0 hours lag), exclusively, and the signs of their correlations with ΔS change in both cases. Owing to the time separation and sign change in each of these variables they are valid predictors, and so equation (11) is also a valid predictor equation for ΔS although somewhat weak in R^2. The remarks made earlier, relative to the signs of the correlations of $-\bar{\alpha}_b$ and $+\bar{D}/\bar{z}$, and the physical effects of these variables at lag time 6·0 hours, are applicable here because the effects are of like kind: i.e., positive ΔS is almost always equivalent to positive ΔQ_f and thus the identity of the signs in Tables 3 and 4 leads to identity in the explanations of the predictor–predictand correlations for $-\bar{\alpha}_b$ and $+\bar{D}/\bar{z}$. The positive correlation between $\bar{\alpha}_b$ and ΔS, at lag 0·0 hours, reflects the fact that there is less erosion of the lower foreshore at low water if the breaker angle is large because runup is not as great then and less material may be scoured away at this time.

The importance of the negative correlation between ΔS and the lead predictor $\bar{D}/\bar{z}$, *at time of low water* (Table 4) appears related to the interplay between groundwater seepage through the saturated foreshore, swash velocities, and the transport of sand grains of various sizes. Thus when pore water pressure at the time of low tide causes maximum lift forces on the sediment grains, large grains may be more easily moved downslope by the backwash than small grains because of the greater flow of the groundwater through the coarse grains. The net motion of coarse grains will be decidedly downslope. For constant values of all other variables, if smaller grains compose the foreshore, there will be a slower flow of groundwater at the free surface (foreshore surface), and grains transported up the foreshore will tend to stay there.

The larger the $\bar{D}/\bar{z}$ ratio (the less the amount of water in the uprush or the greater the friction encountered by the uprush), the smaller will be the *distance* of uprush (b). Under such conditions, grains will not be transported as far up the beach as for a smaller $\bar{D}/\bar{z}$, and downslope movement of grains under the backwash will be more effective, owing to excess pore-water pressure. This negative correlation between the runup b and $\bar{D}/\bar{z}$ is documented further in Table 5, where the results of screening (12) are presented:

$$b = f(\bar{H}_b/g\bar{T}_b^{\,2})^{\frac{1}{2}},\ \rho_l/\rho_s,\ \tan \bar{m},\ \tan \bar{\alpha}_b,\ h,\ \bar{D}/\bar{z} \qquad (12)$$

In screening (12), 96 observations of b were used, one half for low water and one half for high water conditions. The strength of the $\bar{D}/\bar{z}$ correlations with b, in spite of such mixed conditions, suggests that it is of overriding physical importance. This negative correlation may also simply reflect the fact that, for a given beach slope and grain diameter, z increases as b increases.

Table 5 Selection of predictors (for $Y = b$) by screening procedure (half observations = low water; half observations = high water; no lag times; $N = 96$)

Equation	Variable	Sign of correlation	R^2
(1)	$\bar{m}$	−	0·12
(2)	$\bar{D}/\bar{z}$	−	0·20
(3)	h	+	0·25

If erosion of the foreshore and retreat of the shoreline will occur when $\bar{D}/\bar{z}$ (at low water) is large, it must follow that advance of the shoreline will occur when $\bar{D}/\bar{z}$ (at low water) is small; this is in part because the larger values of b at such times will result in the transport of grains to positions higher on the foreshore, where they will remain, being less influenced by lift forces induced by groundwater. Assuming the validity of this physical correlation, we may assume also that (11) will be a useful initial statement for prediction of the advance or retreat of gently sloping sandy beaches on tidal shorelines where the environmental conditions are similar to those at Virginia Beach. Figure 6 is a plot of observed versus predicted values for the data used to develop equation (11).

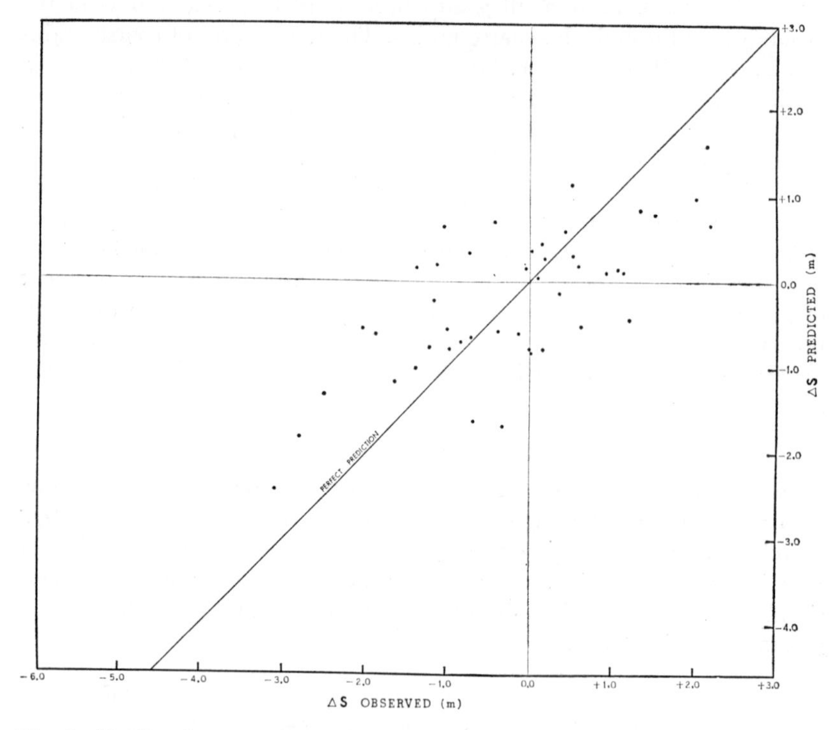

Fig. 6 Predicted versus observed values for advance or retreat of the shoreline, from one low-water stand to the next, using equation (11).

5 Mean slope of the foreshore, $\bar{m}$

In the above work, we have seen the importance to ΔQ_f and ΔS of the interactions between the various factors that shape a beach, including especially the groundwater characteristics. Let us look now at these interactions as they effect that rather elusive element of foreshore geometry—the beach slope.

Suppose we investigate (13) for the low water to low water interval, over 5 observations, each separated by 3 hours. Thus:

$$\bar{m} = f[(\bar{H}b/g\bar{T}_b^2)^{\frac{1}{2}}{}_{0-4}, (\rho_l/\rho_s)_{0-4}, (\bar{\alpha}_b)_{0-4}, (h/b)_{0-4}, (\bar{D}/\bar{z})_{0-4}] \quad (13)$$

The results of this screening run appear in Table 6, where it is seen that h/b is positively correlated with $\bar{m}$ at all lag times except that for falling half tide. Thus as h/b becomes larger, so does $\bar{m}$, except at falling half tide when the slope decreases as the hydraulic head becomes large relative to the swash length, b. The positive correlation between $\bar{m}$ and h/b at 0·0, 6·0, 9·0 and 12·0 hours before the measurement of $\bar{m}$ (at low water) probably reflects the obvious relationship between $\bar{m}$ and $1/b$; that is, smaller breaker distances are to be found on steeper foreshores, and *vice versa*. At time of falling half tide (lag 3·0 hours, Table 6), however, the negative correlation between $\bar{m}$ and h/b is related to the fact that sand is eroded from the upper and middle swash zones and deposited near the breakers when h becomes large relative to b. Observations show that this is almost always the case on a falling tide after a high water stand which has sent swash water above the outcrop of the groundwater table and increased the hydraulic head, h, relative to the head that obtained at rising half tide (lag 9·0 hours). The erosion of material in the upper and middle swash reaches and its deposition in the lower swash zone leads to a lower foreshore slope, as that slope is recorded at the end of the low water interval. It will be recalled that the section of 'the foreshore' used for slope measurement in this study is that segment 'between the inshore margin of the breakers and the top of the swash at a given instant.'

Table 6 Selection of predictors (for $Y = \bar{m}$) by screening procedure (lag interval expressed in hours before low water; $N = 48$)

Equation	Variable	Sign of correlation	Lag (hours)	R^2
(1)	h/b	+	9·0	0·14
(2)	$(\bar{H}_b/g\bar{T}_b^2)^{\frac{1}{2}}$	−	3·0	0·27
(3)	h/b	+	0·0	0·35
(4)	$\bar{\alpha}_b$	+	9·0	0·38
(5)	$\bar{D}/\bar{z}$	−	0·0	0·41
(6)	$\bar{D}/\bar{z}$	+	3·0	0·44
(7)	h/b	−	3·0	0·48
(8)	$\bar{D}/\bar{z}$	−	6·0	0·51
(9)	$\bar{\alpha}_b$	−	3·0	0·53
(10)	h/b	+	12·0	0·56
...	...	...	...	...
(12)	$\bar{D}/\bar{z}$	+	9·0	0·59
...	...	...	...	...
(14)	h/b	+	6·0	0·60

A similar analysis for $\bar{D}/\bar{z}$ begins by noting (Table 6) that this variable is positively correlated with $\bar{m}$ at times of rising or falling half tide, but negatively correlated at time of high water or final low water. A large 'swash depth' ($\bar{z}$) at high water (small $\bar{D}/\bar{z}$ ratio) indicates that there will be a large swash length for a given $\bar{D}$, and this larger swash length will lead to buildup of the foreshore above the outcrop of the groundwater table (u, Figure 4), and a steeper final foreshore slope. So also will a large swash length at final low water transport sand to the mid-foreshore where it will be deposited, resulting in the development of a steeper slope at final low water. At rising or falling half tide, however, the $\bar{m}$—$\bar{D}/\bar{z}$ correlation is reversed and the well-documented relationship between steep foreshore slopes (at mid-tide level) and larger diameter sand grains—or *vice versa*—is echoed by the results of the regression analysis.

The strong negative correlation between $(\bar{H}_b/g\bar{T}_b{}^2)^{\frac{1}{2}}{}_{3\cdot0}$ and $\bar{m}$ (Table 6)

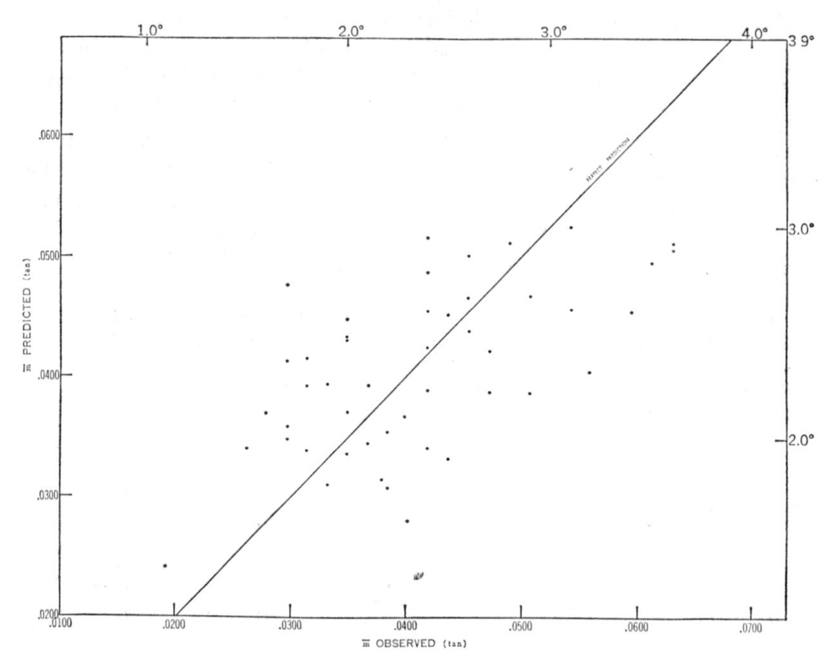

Fig. 7 Predicted versus observed values for slope of the foreshore, as measured at the end of the low-water interval, using equation (14).

is found also in Table 4, where ΔC is the predictand, and to a lesser extent (lower R^2) in Table 3, where ΔQ_f is the predictand. Thus the multi-regression analysis indicates that the foreshore at the end of a low water interval has been shaped by breaker steepness at falling half tide in such a way that, as breaker steepness increases at lag 3·0 hours:

1 there tends to be net erosion of material from the foreshore
2 the mean high water line (shoreline) retreats
3 the foreshore slope—between the inshore margin of the breakers and the top of the swash at low water—becomes lower.

The predictor equation for $\bar{m}$, containing the largest number of predictors that is feasible, is given below:

$$\bar{m} = 0{\cdot}345 + 0{\cdot}351\ (h/b)_{9\cdot0} - 0{\cdot}374(\bar{H}_b/g\bar{T}_b^{\,2})^{\frac{1}{2}}{}_{3\cdot0} + 0{\cdot}238(h/b)_{0\cdot0}$$
$$+ 0{\cdot}032(\bar{\alpha}_b)_{9\cdot0} - 0{\cdot}011(\bar{D}/\bar{z})_{0\cdot0} + 0{\cdot}018(\bar{D}/\bar{z})_{3\cdot0} - 0{\cdot}217(/b)_{3\cdot0} \quad (14)$$

where $\bar{m}$ is the tangent of the mean slope angle. Figure 7 is a plot of observed versus predicted values for the data used in developing (14).

6 Alternative variables and the best predictor equations

In working with correlation techniques such as linear multiregression, one is always concerned as to whether one has formulated the most powerful independent variables from one's data. For example, although the index for breaker steepness $(\bar{H}_b/g\bar{T}_b^{\,2})^{\frac{1}{2}}$ used in the foregoing gives a relatively adequate description of the breaker *form*, it does not take into account the relative *sizes* of the breakers. Thus at different times one could observe breakers of similar steepness but of widely divergent energy content, owing to breaker height differences. What if an index for breaker power were used? Would it be a more powerful predictor of the Y_i, either alone or in combination with the other X_i?

With such a consideration in mind, an index of breaker 'power' was substituted for the breaker steepness index. The breaker power index used is simply an estimate of the total energy contained in a breaking wave of average height $(\bar{H}_b)$ divided by the mean breaker period $(\bar{T}_b)$; that is, the index is a representation of the breaker energy transmission (power).

The total energy of the average breaker $\bar{E}_{bt}$ was approximated by the total wave energy (cf. Munk, 1949, equation 13) for the average solitary wave at breaking, $\bar{E}_t$:

$$\bar{E}_{bt} \approx \bar{E}_t = \frac{8}{3\sqrt{3}}\rho g\, H^{\frac{3}{2}}\, h^{\frac{3}{2}} \text{ (in ft lbs/ft of crest width)}$$

where H is approximated by $\bar{H}_b$ and h by $\bar{z}$. Thus the estimate of mean breaker power $\bar{P}_b$ becomes:

$$\bar{P}_b \approx \frac{8/3\sqrt{3}\ \rho g\ \bar{H}_b^{\frac{3}{2}}\ \bar{z}^{\frac{3}{2}}}{\bar{T}_b} \cdot 0{\cdot}1383 \text{ (in kg m per m of crest width)}$$

The results of screening equation (15) are presented in Table 7, for comparison with the results (Table 6) of screening equation (13).

$$\bar{m} = f[(\bar{P}_b)_{0-4}, (\rho_l/\rho_s)_{0-4}, (\bar{\alpha}_b)_{0-4}, (h/b)_{0-4}, (\bar{D}/\bar{z})_{0-4}] \quad (15)$$

It is seen that $(\bar{P}_b)_{3\cdot0}$ was indeed selected by the screening process as the strongest single predictor of $\bar{m}$, leading $(h/b)_{9\cdot0}$ by 0·05 in R^2 (cf.

Tables 7 and 6 respectively). When in combination with two or more

Table 7 Selection of predictors by screening equation (15)

Equation	Variable	Sign of correlation	Lag (hours)	R^2
(1)	$\bar{P}_b$	−	3·0	0·19
(2)	h/b	+	9·0	0·28
(3)	h/b	+	0·0	0·33
(4)	$\bar{D}/\bar{z}$	−	0·0	0·38
(5)	$\bar{\alpha}_b$	+	9·0	0·42
(6)	h/b	−	3·0	0·45
(7)	$\bar{P}_b$	+	6·0	0·48

X_i, however, the substitution of $\bar{P}_b$ for $(\bar{H}_b/g\bar{T}_b^2)^{\frac{1}{2}}$ results in nearly identical R^2 values for X_i from 2 to 7. Thus while it is interesting to note the initial strength of $\bar{P}_b$ relative to the other independent variables, its strength makes little difference in a *multi*predictor equation containing two or more X_i. Equation (16), then, predicts beach slope as well as equation (14).

$$\bar{m} = 0{\cdot}039 - 0{\cdot}0002(\bar{P}_b)_{3\cdot0} + 0{\cdot}415(h/b)_{9\cdot0} + 0{\cdot}242(h/b)_{0\cdot0} - 0{\cdot}014(\bar{D}/\bar{z})_{0\cdot0} + 0{\cdot}032(\bar{\alpha}_b)_{9\cdot0} - 0{\cdot}169(h/b)_{3\cdot0} + 0{\cdot}00004(\bar{E}_b)_{6\cdot0} \quad (16)$$

III Evaluation

In the foregoing attempt to develop quantitative expressions for geometrical changes in a tidal beach, a statistical procedure has been adopted which involves linear correlations between dependent and independent variables. The linear model may not be the best one. In this attempt to develop predictor equations for geometries developed over a tidal-cycle interval, however, the model has paid reasonable dividends.

The limitations of multiregression analysis of dimensionally homogeneous expressions have been shown, and the advantage of experimenting instead with intuitively meaningful dimensionless variables (X_i) has been demonstrated.

The important effects of the groundwater flow and the relation of the swash geometry to the outcrop of the groundwater table on changes in beach geometry are particularly well brought out by the multiregression analyses of this study. No laboratory experiment on beach formation will adequately model a natural beach unless provisions for simulation of the water table characteristics are included. (Bagnold [1940, 42] recognised the importance of swash percolation in laboratory experiments nearly 30 years ago.) Further development of prediction techniques for changes in natural beaches will have to include pertinent data for the groundwater characteristics.

Equations (9), (11), (14) and (16) should be tested using independent data gathered from similar beach environments. At the moment the equations appear to be the best statements available of the dependency of ΔQ_f, ΔS, and $\bar{m}$ on environmental variables that express themselves on the gently sloping beach of the mid-Atlantic coast which was investigated here.

IV Summary and conclusion

The prediction of beach changes is hampered by the following aspects of the beach–ocean–atmosphere system:

1 *Short memory*—the 'process' variables are constantly changing so that in nearly all cases the magnitude of a given process variable changes before a full beach response is realised
2 *Feedback*—the 'independent' variables in the system are interlocked to various degrees (*i.e.*, 'interdependent'), so that a change in one process variable induces changes in other process variables
3 *Range in values of process variables*—the process variables may fluctuate in frequency, magnitude and duration through large ranges, and thus the number of combinations of such fluctuating variables, each combination of which may produce a unique beach response, is very large.

An approach to correlation and prediction in the beach–ocean–atmosphere system has been presented. It involves the application of linear multi-regression analysis to time-series data from the beach environment. An example has been given of the application of the approach in which the following geometrical changes were investigated for a gently sloping beach at Virginia Beach over the interval from one low water to the next:

1 net change in quantity of foreshore sand ΔQ_f (cross-sectional area of change in a plane perpendicular to shore, times a unit distance parallel to shore),
2 advance or retreat of the shoreline (mean high water line) ΔS, and
3 mean slope of the foreshore, $\bar{m}$.

The data for ΔQ_f, ΔS, and $\bar{m}$ were derived from 48 consecutive low water profiles made along a single transect. 15 environmental variables were also measured, concurrent with the monitoring of foreshore changes. Linear multiregression analyses performed on the data resulted in useful prediction equations of moderate strength.

Owing to the complexly interlocked nature of the variables in the beach–ocean–atmosphere system, it would seem that the best immediate prospect for valid predictions of changes in beach geometry is to be found in empirical equations based upon the correlation of reliable time-series data from various beach environments. Although the linear multiregression technique of correlation used in this and previous studies shows promise, more powerful techniques of time-series analysis, such as spectral analysis, should be employed in the search for optimum empirical equations.

V Acknowledgements

N. A. Pore, Weather Bureau, ESSA, programmed the screening run. I thank R. J. Byrne, P. H. Kemp and W. C. Krumbein for critical review of portions of the manuscript.

VI References

Adams, K. T. 1942: Hydrographic manual. *United States Department of Commerce, Special Publication* 143, Washington D.C.: Government Printer. (940 pp.)

Bagnold, R. A. 1940: Beach formation by waves: some model experiments in a wave tank. *Journal of the Institute of Civil Engineers* 15, 27–52.

Benson, M. A. 1965: Spurious correlation in hydraulics and hydrology. *American Society of Civil Engineers, Journal of the Hydraulics Division* 91, 35–41.

Dolan, R. 1965: *Relationships between nearshore processes and beach changes along the Outer Banks of North Carolina*. Louisiana State University: PhD thesis. (51 pp.)

Duncan, J. R. Jr 1964: The effects of water table and tide cycle on swash-backwash sediment distribution and beach profile development. *Marine Geology* 2, 186–97.

Galvin, C. J. Jr 1968: Breaker type classification on three laboratory beaches. *Journal of Geophysical Research* 73, 3651–9.

Giese, G. S. 1966 : *Beach pebble movements and shape sorting indices of swash zone mechanics.* University of Chicago : PhD thesis. (65 pp.)

Harrison, W. and **Krumbein, W. C.** 1964 : Interactions of the beach–ocean atmosphere system at Virginia Beach, Virginia. *United States Army Corps of Engineers, Coastal Engineering Research Centre, Technical Memo* 7. (102 pp.)

Harrison, W. and **Pore, N. A.** 1964 : Alternative multiregression technique for obtaining predictor equations. *Addendum to CERC Technical Memo* 7 (Harrison and Krumbein, 1964). (8 pp.)

1967 : An approach to correlation and prediction in the wind–runoff–drift system. In Circulation of shelf waters off the Chesapeake Bight, *Environmental Science Services Administration, Professional Paper* 3, 43–78. (82 pp.)

Harrison, W., Pore, N. A. and **Tuck, R. D.** 1965 : Predictor equations for beach processes and responses. *Journal of Geophysical Research* 70, 6103–9.

Harrison, W., Rayfield, E. W., Boon, J. D., Reynolds, G., Grant, J. B. and **Tyler, D.** 1968 : A time series from the beach environment. *Environmental Science Services Administration, Research Laboratories, Technical Memo* AOL-1. (85 pp.)

Inman, D. L. and **Rusnak, G. A.** 1956 : Changes in sand level on the beach and shelf at La Jolla, California, *United States Army, Corps of Engineers, Beach Erosion Board, Technical Memo* 82. (32 pp.)

Kemp, P. H. 1961 : The relationship between wave action and beach profile characteristics. *Proceedings of the 7th Conference on Coastal Engineering*, New York : American Society of Engineers, 262–77.

Miller, R. G. 1958 : The screening procedure. *Hartford, Connecticut, Travelers Weather Research Center, Studies in Statistical Weather Prediction, Final Report, Contract* AF19(604)–1590, 86–95.

Munk, W. H. 1949 : The solitary wave and its application to surf problems. *Annals of the New York Academy of Science* 51, 376–424.

Sonu, C. J. and **Russell, R. J.** 1967 : Topographic changes in the surf zone profile. *Proceedings of the 10th Conference of Coastal Engineering*, New York : American Society of Engineers, 525–49.

Strahler, A. N. 1964 : Tidal cycle of changes in an equilibrium beach, Sandy Hook, New Jersey. *Columbia University, Department of Geology, Technical Report* 4, *project* NR388–057, *Contract* NONR 266(68). (28 pp.)